Applied Generative AI for Beginners

Practical Knowledge on Diffusion Models, ChatGPT, and Other LLMs

Akshay Kulkarni
Adarsha Shivananda
Anoosh Kulkarni
Dilip Gudivada

Apress®

Applied Generative AI for Beginners: Practical Knowledge on Diffusion Models, ChatGPT, and Other LLMs

Akshay Kulkarni
Bangalore, Karnataka, India

Anoosh Kulkarni
Bangalore, Karnataka, India

Adarsha Shivananda
Hosanagara, Karnataka, India

Dilip Gudivada
Bangalore, India

ISBN-13 (pbk): 978-1-4842-9993-7
https://doi.org/10.1007/978-1-4842-9994-4

ISBN-13 (electronic): 978-1-4842-9994-4

Managing Director, Apress Media LLC: Welmoed Spahr
Acquisitions Editor: Celestin Suresh John
Development Editor: Laura Berendson
Editorial Assistant: Gryffin Winkler

Cover designed by eStudioCalamar

Cover image designed by Scott Webb on unsplash

Distributed to the book trade worldwide by Springer Science+Business Media New York, 1 New York Plaza, Suite 4600, New York, NY 10004-1562, USA. Phone 1-800-SPRINGER, fax (201) 348-4505, e-mail orders-ny@ springer-sbm.com, or visit www.springeronline.com. Apress Media, LLC is a California LLC and the sole member (owner) is Springer Science + Business Media Finance Inc (SSBM Finance Inc). SSBM Finance Inc is a **Delaware** corporation.

For information on translations, please e-mail booktranslations@springernature.com; for reprint, paperback, or audio rights, please e-mail bookpermissions@springernature.com.

Apress titles may be purchased in bulk for academic, corporate, or promotional use. eBook versions and licenses are also available for most titles. For more information, reference our Print and eBook Bulk Sales web page at http://www.apress.com/bulk-sales.

Any source code or other supplementary material referenced by the author in this book is available to readers on GitHub. For more detailed information, please visit https://www.apress.com/gp/services/ source-code.

Paper in this product is recyclable

To our families

Table of Contents

About the Authors

Akshay Kulkarni is an AI and machine learning evangelist and IT leader. He has assisted numerous Fortune 500 and global firms in advancing strategic transformations using AI and data science. He is a Google Developer Expert, author, and regular speaker at major AI and data science conferences (including Strata, O'Reilly AI Conf, and GIDS). He is also a visiting faculty member for some of the top graduate institutes in India. In 2019, he was featured as one of the top 40 under-40 data scientists in India. He enjoys reading, writing, coding, and building next-gen AI products.

Adarsha Shivananda is a data science and generative AI leader. Presently, he is focused on creating world-class MLOps and LLMOps capabilities to ensure continuous value delivery using AI. He aims to build a pool of exceptional data scientists within and outside the organization to solve problems through training programs and always wants to stay ahead of the curve. He has worked in the pharma, healthcare, CPG, retail, and marketing industries. He lives in Bangalore and loves to read and teach data science.

Anoosh Kulkarni is a data scientist and MLOps engineer. He has worked with various global enterprises across multiple domains solving their business problems using machine learning and AI. He has worked at one of the leading ecommerce giants in UAE, where he focused on building state-of-the-art recommender systems and deep learning–based search engines. He is passionate about guiding and mentoring people in their data science journey. He often leads data science/machine learning meetups, helping aspiring data scientists carve their career road map.

Dilip Gudivada is a seasoned senior data architect with 13 years of experience in cloud services, big data, and data engineering. Dilip has a strong background in designing and developing ETL solutions, focusing specifically on building robust data lakes on the Azure cloud platform. Leveraging technologies such as Azure Databricks, Data Factory, Data Lake Storage, PySpark, Synapse, and Log Analytics, Dilip has helped organizations establish scalable and efficient data lake solutions on Azure. He has a deep understanding of cloud services and a track record of delivering successful data engineering projects.

About the Technical Reviewer

Prajwal is a lead applied scientist and consultant in the field of generative AI. He is passionate about building AI applications in the service of humanity.

Introduction

Welcome to *Applied Generative AI for Beginners: Practical Knowledge on Diffusion Models, ChatGPT, and Other LLMs*. Within these pages, you're about to embark on an exhilarating journey into the world of generative artificial intelligence (AI). This book serves as a comprehensive guide that not only unveils the intricacies of generative AI but also equips you with the knowledge and skills to implement it.

In recent years, generative AI has emerged as a powerhouse of innovation, reshaping the technological landscape and redefining the boundaries of what machines can achieve. At its core, generative AI empowers artificial systems to understand and generate human language with remarkable fluency and creativity. As we delve deep into this captivating landscape, you'll gain both a theoretical foundation and practical insights into this cutting-edge field.

What You Will Discover

Throughout the chapters of this book, you will

- Build Strong Foundations: Develop a solid understanding of the core principles that drive generative AI's capabilities, enabling you to grasp its inner workings.

- Explore Cutting-Edge Architectures: Examine the architecture of large language models (LLMs) and transformers, including renowned models like ChatGPT and Google Bard, to understand how these models have revolutionized AI.

- Master Practical Implementations: Acquire hands-on skills for integrating generative AI into your projects, with a focus on enterprise-grade solutions and fine-tuning techniques that enable you to tailor AI to your specific needs.

- Operate with Excellence: Discover LLMOps, the operational backbone of managing generative AI models, ensuring efficiency, reliability, and security in your AI deployments.

- Witness Real-World Use Cases: Explore how generative AI is revolutionizing diverse domains, from business and healthcare to creative writing and legal compliance, through a rich tapestry of real-world use cases.

CHAPTER 1

Introduction to Generative AI

Have you ever imagined that simply by picturing something and typing, an image or video could be generated? How fascinating is that? This concept, once relegated to the realm of science fiction, has become a tangible reality in our modern world. The idea that our thoughts and words can be transformed into visual content is not only captivating but a testament to human innovation and creativity.

Figure 1-1. *The machine-generated image based on text input*

Even as data scientists, many of us never anticipated that AI could reach a point where it could generate text for a specific use case. The struggles we faced in writing code or the countless hours spent searching on Google for the right solution were once common challenges. Yet, the technological landscape has shifted dramatically, and those laborious tasks have become relics of the past.

© Akshay Kulkarni, Adarsha Shivananda, Anoosh Kulkarni, Dilip Gudivada 2023
A. Kulkarni et al., *Applied Generative AI for Beginners*, https://doi.org/10.1007/978-1-4842-9994-4_1

How has this become possible? The answer lies in the groundbreaking advancements in deep learning and natural language processing (NLP). These technological leaps have paved the way for generative AI, a field that harnesses the power of algorithms to translate thoughts into visual representations or automates the creation of complex code. Thanks to these developments, we're now experiencing a future where imagination and innovation intertwine, transforming the once-unthinkable into everyday reality.

So, What Is Generative AI?

Generative AI refers to a branch of artificial intelligence that focuses on creating models and algorithms capable of generating new, original content, such as images, text, music, and even videos. Unlike traditional AI models that are trained to perform specific tasks, generative AI models aim to learn and mimic patterns from existing data to generate new, unique outputs.

Generative AI has a wide range of applications. For instance, in computer vision, generative models can generate realistic images, create variations of existing images, or even complete missing parts of an image. In natural language processing, generative models can be used for language translation, text synthesis, or even to create conversational agents that produce humanlike responses. Beyond these examples, generative ai can perform art generation, data augmentation, and even generating synthetic medical images for research and diagnosis. It's a powerful and creative tool that allows us to explore the boundaries of what's possible in computer vision.

However, it's worth noting that generative AI also raises ethical concerns. The ability to generate realistic and convincing fake content can be misused for malicious purposes, such as creating deepfakes or spreading disinformation. As a result, there is ongoing research and development of techniques to detect and mitigate the potential negative impacts of generative AI.

Overall, generative AI holds great promise for various creative, practical applications and for generating new and unique content. It continues to be an active area of research and development, pushing the boundaries of what machines can create and augmenting human creativity in new and exciting ways.

Components of AI

- Artificial Intelligence (AI): It is the broader discipline of machine learning to perform tasks that would typically require human intelligence.

- Machine Learning (ML): A subset of AI, ML involves algorithms that allow computers to learn from data rather than being explicitly programmed to do so.

- Deep Learning (DL): A specialized subset of ML, deep learning involves neural networks with three or more layers that can analyze various factors of a dataset.

- Generative AI: An advanced subset of AI and DL, generative AI focuses on creating new and unique outputs. It goes beyond the scope of simply analyzing data to making new creations based on learned patterns.

Figure 1-2 explains how generative AI is a component of AI.

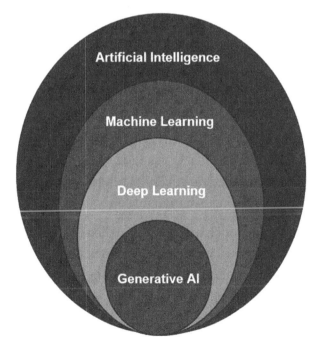

Figure 1-2. *AI and its components*

Domains of Generative AI

Let's deep dive into domains of generative AI in detail, including what it is, how it works, and some practical applications.

Text Generation

- What It Is: Text generation involves using AI models to create humanlike text based on input prompts.

- How It Works: Models like GPT-3 use Transformer architectures. They're pre-trained on vast text datasets to learn grammar, context, and semantics. Given a prompt, they predict the next word or phrase based on patterns they've learned.

- Applications: Text generation is applied in content creation, chatbots, and code generation. Businesses can use it for crafting blog posts, automating customer support responses, and even generating code snippets. Strategic thinkers can harness it to quickly draft marketing copy or create personalized messages for customers.

Image Generation

- What It Is: Image generation involves using various deep learning models to create images that look real.

- How It Works: GANs consist of a generator (creates images) and a discriminator (determines real vs. fake). They compete in a feedback loop, with the generator getting better at producing images that the discriminator can't distinguish from real ones.

- Applications: These models are used in art, design, and product visualization. Businesses can generate product mock-ups for advertising, create unique artwork for branding, or even generate faces for diverse marketing materials.

Audio Generation

- What It Is: Audio generation involves AI creating music, sounds, or even humanlike voices.

- How It Works: Models like WaveGAN analyze and mimic audio waveforms. Text-to-speech models like Tacotron 2 use input text to generate speech. They're trained on large datasets to capture nuances of sound.

- Applications: AI-generated music can be used in ads, videos, or as background tracks. Brands can create catchy jingles or custom sound effects for marketing campaigns. Text-to-speech technology can automate voiceovers for ads or customer service interactions. Strategically, businesses can use AI-generated audio to enhance brand recognition and storytelling.

Video Generation

- What It Is: Video generation involves AI creating videos, often by combining existing visuals or completing missing parts.

- How It Works: Video generation is complex due to the temporal nature of videos. Some models use text descriptions to generate scenes, while others predict missing frames in videos.

- Applications: AI-generated videos can be used in personalized messages, dynamic ads, or even content marketing. Brands can craft unique video advertisements tailored to specific customer segments. Thoughtful application can lead to efficient video content creation that adapts to marketing trends.

Generating Images

Microsoft Bing Image Creator is a generative AI tool that uses artificial intelligence to create images based on your text descriptions.

www.bing.com/images/create/

To use Bing Image Creator, you simply type a description of the image you want to create into the text box. We will use the same example mentioned earlier in **generating realistic images.** "Create an image of a pink elephant wearing a party hat and standing on a rainbow." Bing Image Creator will then generate an image based on your description.

Figure 1-3 shows the Microsoft Bing output.

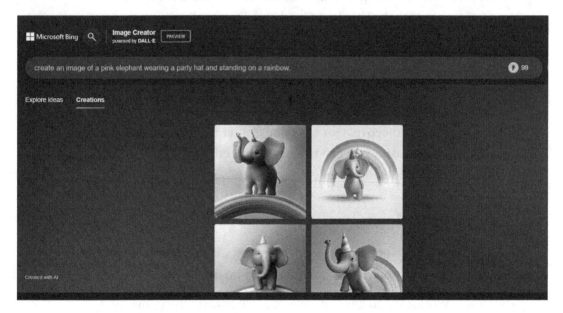

Figure 1-3. *Microsoft Bing output*

Generating Text

Let's use ChatGPT for generating text. It is a large language model–based chatbot developed by OpenAI and launched in November 2022.

ChatGPT is trained with reinforcement learning through human feedback and reward models that rank the best responses. This feedback helps augment ChatGPT with machine learning to improve future responses.

ChatGPT can be used for a variety of purposes, including

- Having conversations with users

- Answering questions

- Generating text

- Translating languages

- Writing different kinds of creative content

ChatGPT can be accessed online at
`https://openai.com/blog/chatgpt`
To use ChatGPT, you simply type a description you want into the text box.
To create content on our solar system. Figure 1-4 shows the ChatGPT's output.

 create a content on solar system

 The Wonders of Our Solar System: A Journey Through Space

The Solar System, a captivating cosmic dance of planets, moons, asteroids, and comets, is a majestic spectacle in the vast expanse of space. Our home planet, Earth, is just one of the many celestial bodies that make up this extraordinary cosmic neighborhood. So, let's embark on a journey through the wonders of our Solar System and explore the secrets it holds!

1. The Sun - Our Life-Giving Star:
At the heart of the Solar System shines the Sun, a brilliant ball of hot gas that provides light and warmth to all the planets. Its colossal energy sustains life on Earth and fuels the intricate dance of celestial bodies around it. The Sun's fiery surface is a mesmerizing dance of solar flares and sunspots, revealing its ever-changing nature.

2. Mercury - The Swiftest Planet:
Moving closer to the Sun, we encounter Mercury, the closest planet to our star. This rocky, desolate world is a scorching hellscape during the day and a freezing tundra at night. It speeds around the Sun, completing an orbit in just 88 Earth days, making it the swiftest planet in the Solar System.

3. Venus - The Enigmatic Beauty:
Next, we encounter Venus, often called Earth's sister planet due to its similar size and composition. However, its atmosphere is a toxic mix of sulfuric acid and suffocating carbon dioxide, creating a runaway greenhouse effect. Venus boasts a thick layer of clouds that veils its surface, making it one of the most enigmatic and mysterious planets in our Solar System.

4. Earth - Our Blue Marble:
Our home planet, Earth, stands out as a vibrant oasis in the darkness of space. The only known planet to support life, Earth is adorned with lush forests, vast oceans, and a diverse array of life forms. Its intricate climate systems and ecosystems make it a haven for living beings to thrive.

Figure 1-4. *ChatGPT's output*

5. Mars - The Red Planet:

Moving outward, we encounter the captivating Mars, a rust-colored world with a thin atmosphere. Mars has long been the focus of scientific exploration, with its dusty plains, towering volcanoes, and the possibility of ancient water channels hinting at past signs of life.

6. Jupiter - The Giant Among Giants:

Entering the realm of the gas giants, we meet Jupiter, the largest planet in the Solar System. Its mesmerizing bands of clouds, raging storms (like the iconic Great Red Spot), and numerous moons, including the four Galilean moons, make it a celestial wonder to behold.

7. Saturn - The Ringed Beauty:

Saturn's majestic rings, made of icy particles, are an awe-inspiring sight. This gas giant's golden hue and graceful ring system make it a true wonder of the Solar System. Its moon Titan, with its dense atmosphere and liquid methane lakes, remains one of the most Earth-like worlds beyond our home planet.

8. Uranus and Neptune - The Ice Giants:

The final two gas giants, Uranus and Neptune, are known as the ice giants due to their significant icy and rocky components. Neptune's vibrant blue color and its intriguing moon Triton, with geysers erupting from its surface, add to the allure of these distant worlds.

9. Pluto and the Kuiper Belt:

As we venture beyond Neptune, we reach the icy realm of the Kuiper Belt, home to countless icy bodies and dwarf planets. Among them is Pluto, once considered the ninth planet, now reclassified as a dwarf planet. Pluto's fascinating system of moons and icy plains make it a captivating world in its own right.

10. Moons, Comets, and Asteroids:

Throughout our journey, we encounter a myriad of moons, comets, and asteroids. From Jupiter's moon Europa, with its subsurface ocean, to the icy geysers of Enceladus around Saturn, and the awe-inspiring comets that streak through the sky, these celestial companions add richness to the tapestry of the Solar System.

The Solar System, with its diverse and captivating array of celestial bodies, continues to inspire exploration and discovery. As we learn more about our cosmic neighborhood, the wonders of the Solar System remind us of the vastness and beauty of the universe, beckoning us to continue our exploration beyond the stars.

Figure 1-4. *(continued)*

ChatGPT or any other tools are still under development, but it has learned to perform many kinds of tasks. As it continues to learn, it will become even more powerful and versatile.

Generative AI: Current Players and Their Models

Generative AI is a rapidly growing field with the potential to revolutionize many industries. Figure 1-5 shows some of the current players in the generative AI space.

Figure 1-5. *ChatGPT's output*

Briefly let's discuss few of them:

- OpenAI: OpenAI is a generative AI research company that was founded by Elon Musk, Sam Altman, and others. OpenAI has developed some of the most advanced generative AI models in the world, including GPT-4 and DALL-E 2.

  - GPT-4: GPT-4 is a large language model that can generate text, translate languages, write different kinds of creative content, and answer your questions in an informative way.

  - DALL-E 2: DALL-E 2 is a generative AI model that can create realistic images from text descriptions.

- DeepMind: DeepMind is a British artificial intelligence company that was acquired by Google in 2014. DeepMind has developed several generative AI models, including AlphaFold, which can predict the structure of proteins, and Gato, which can perform a variety of tasks, including playing Atari games, controlling robotic arms, and writing different kinds of creative content.

- Anthropic: Anthropic is a company that is developing generative AI models for use in a variety of industries, including healthcare, finance, and manufacturing. Anthropic's models are trained on massive datasets of real-world data, which allows them to generate realistic and accurate outputs.

- Synthesia: Synthesia is a company that specializes in creating realistic synthetic media, such as videos and audio recordings. Synthesia's technology can be used to create avatars that can speak, gesture, and even lip-sync to any audio input.

  - RealSpeaker: RealSpeaker is a generative AI model that can be used to create realistic synthetic voices.

  - Natural Video: Natural Video is a generative AI model that can be used to create realistic synthetic videos.

- RunwayML: RunwayML is a platform that makes it easy for businesses to build and deploy generative AI models. RunwayML provides a variety of tools and resources to help businesses collect data, train models, and evaluate results.

  - Runway Studio: Runway Studio is a cloud-based platform that allows businesses to build and deploy generative AI models without any coding experience.

  - Runway API: The Runway API is a set of APIs that allow businesses to integrate generative AI into their applications.

- Midjourney: Midjourney is a generative AI model that can be used to create realistic images, videos, and text. Midjourney is still under development, but it has already been used to create some impressive results.

These are just a few of the many companies that are working on generative AI. As the field continues to develop, we can expect to see even more innovation and disruption in the years to come.

Generative AI Applications

Generative AI offers a wide array of applications across various industries. Here are some key applications:

1. Content Creation:

   - Text Generation: Automating blog posts, social media updates, and articles.

   - Image Generation: Creating custom visuals for marketing campaigns and advertisements.

   - Video Generation: Crafting personalized video messages and dynamic ads.

2. Design and Creativity:

   - Art Generation: Creating unique artworks, illustrations, and designs.

   - Fashion Design: Designing clothing patterns and accessories.

   - Product Design: Generating prototypes and mock-ups.

3. Entertainment and Media:

   - Music Composition: Creating original music tracks and soundscapes.

   - Film and Animation: Designing characters, scenes, and animations.

   - Storytelling: Developing interactive narratives and plotlines.

4. Marketing and Advertising:

   - Personalization: Crafting tailored messages and recommendations for customers.

   - Branding: Designing logos, packaging, and visual identity elements.

   - Ad Campaigns: Developing dynamic and engaging advertisements.

5. Gaming:

- World Building: Generating game environments, terrains, and landscapes.

- Character Design: Creating diverse and unique in-game characters.

- Procedural Content: Generating levels, quests, and challenges.

6. Healthcare and Medicine:

- Drug Discovery: Designing new molecules and compounds.

- Medical Imaging: Enhancing and reconstructing medical images.

- Personalized Medicine: Tailoring treatment plans based on patient data.

7. Language Translation:

- Real-time Translation: Enabling instant translation of spoken or written language.

- Subtitling and Localization: Automatically generating subtitles for videos.

8. Customer Service:

- Chatbots: Creating conversational agents for customer support.

- Voice Assistants: Providing voice-based assistance for inquiries and tasks.

9. Education and Training:

- Interactive Learning: Developing adaptive learning materials.

- Simulations: Creating realistic training scenarios and simulations.

10. Architecture and Design:

- Building Design: Generating architectural layouts and designs.

- Urban Planning: Designing cityscapes and urban layouts.

Conclusion

This chapter focused on generative AI, a rapidly evolving domain in artificial intelligence that specializes in creating new, unique content such as text, images, audio, and videos. Built upon advancements in deep learning and natural language processing (NLP), these models have various applications, including content creation, design, entertainment, healthcare, and customer service. Notably, generative AI also brings ethical concerns, particularly in creating deepfakes or spreading disinformation. The chapter provides an in-depth look at different domains of generative AI—text, image, audio, and video generation—detailing how they work and their practical applications. It also discusses some of the key players in the industry, like OpenAI, DeepMind, and Synthesia, among others. Lastly, it outlines a wide array of applications across various industries.

CHAPTER 2

Evolution of Neural Networks to Large Language Models

Over the past few decades, language models have undergone significant advancements. Initially, basic language models were employed for tasks such as speech recognition, machine translation, and information retrieval. These early models were constructed using statistical methods, like n-gram and hidden Markov models. Despite their utility, these models had limitations in terms of accuracy and scalability.

With the introduction of deep learning, neural networks became more popular for language modeling tasks. Among them, recurrent neural networks (RNNs) and long short-term memory (LSTM) networks emerged as particularly effective choices. These models excel at capturing sequential relationships in linguistic data and generating coherent output.

In recent times, attention-based approaches, exemplified by the Transformer architecture, have gained considerable attention. These models produce output by focusing on specific segments of the input sequence, using self-attention techniques. Their success has been demonstrated across various natural language processing tasks, including language modeling.

Figure 2-1 shows the key milestones and advancements in the evolution of language models.

© Akshay Kulkarni, Adarsha Shivananda, Anoosh Kulkarni, Dilip Gudivada 2023
A. Kulkarni et al., *Applied Generative AI for Beginners*, https://doi.org/10.1007/978-1-4842-9994-4_2

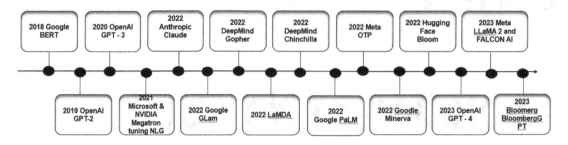

Figure 2-1. *Evolution of language models*

Before hopping to the evolution in detail, let's explore natural language processing.

Natural Language Processing

Natural language processing (NLP) is a subfield of artificial intelligence (AI) and computational linguistics that focuses on enabling computers to understand, interpret, and generate human language. NLP aims to bridge the gap between human communication and machine understanding, allowing computers to process and derive meaning from textual data. It plays a crucial role in various applications, including language translation, sentiment analysis, chatbots, voice assistants, text summarization, and more.

Recent advancements in NLP have been driven by deep learning techniques, especially using Transformer-based models like BERT (Bidirectional Encoder Representations from Transformers) and GPT (Generative Pre-trained Transformer). These models leverage large-scale pre-training on vast amounts of text data and can be fine-tuned for specific NLP tasks, achieving state-of-the-art performance across a wide range of applications.

NLP continues to be a rapidly evolving field, with ongoing research and development aiming to enhance language understanding, generation, and interaction between machines and humans. As NLP capabilities improve, it has the potential to revolutionize the way we interact with technology and enable more natural and seamless human–computer communication.

Tokenization

Tokenization is the process of breaking down the text into individual words or tokens. It helps in segmenting the text and analyzing it at a more granular level.

Example:

Input: "I Love to code in python"

Tokenization: ["I", "Love", "to", "code", "in", "python"]

N-grams

In natural language processing (NLP), n-grams are a powerful and widely used technique for extracting contextual information from text data. N-grams are essentially contiguous sequences of n items, where the items can be words, characters, or even phonemes, depending on the context. The value of "n" in n-grams determines the number of consecutive items in the sequence. Commonly used n-grams include unigrams (1-grams), bigrams (2-grams), trigrams (3-grams), and so on:

1. Unigrams (1-grams):

   Unigrams are single words in a text. They represent individual tokens or units of meaning in the text.

   Example:

   Input: "I love natural language processing."

   Unigrams: ["I", "love", "natural", "language", "processing", "."]

2. Bigrams (2-grams):

   Bigrams consist of two consecutive words in a text. They provide a sense of word pairs and the relationship between adjacent words.

   Example:

   Input: "I love natural language processing."

   Bigrams: [("I", "love"), ("love", "natural"), ("natural", "language"), ("language", "processing"), ("processing", ".")]

3. Trigrams (3-grams):

Trigrams are three consecutive words in a text. They capture more context and provide insights into word triplets.

Example:

Input: "I love natural language processing."

Trigrams: [("I", "love", "natural"), ("love", "natural", "language"), ("natural", "language", "processing"), ("language", "processing", ".")]

4. N-grams in Language Modeling:

In language modeling tasks, n-grams are used to estimate the probability of a word given its context. For example, with bigrams, we can estimate the likelihood of a word based on the preceding word.

5. N-grams in Text Classification:

N-grams are useful in text classification tasks, such as sentiment analysis. By considering the frequencies of n-grams in positive and negative texts, the classifier can learn the distinguishing features of each class.

6. Limitations of n-grams:

While n-grams are powerful in capturing local context, they may lose global context. For instance, bigrams may not be sufficient to understand the meaning of a sentence if some words have strong dependencies on others located farther away.

7. Handling Out-of-Vocabulary (OOV) Words:

When using n-grams, it's essential to handle out-of-vocabulary words (words not seen during training). Techniques like adding a special token for unknown words or using character-level n-grams can be employed.

8. Smoothing:

N-gram models may suffer from data sparsity, especially when dealing with higher-order n-grams. Smoothing techniques like Laplace (add-one) smoothing or Good-Turing smoothing can help address this issue.

N-grams are a valuable tool in NLP for capturing local context and extracting meaningful features from text data. They have various applications in language modeling, text classification, information retrieval, and more. While n-grams provide valuable insights into the structure and context of text, they should be used in conjunction with other NLP techniques to build robust and accurate models.

Language Representation and Embeddings

Language representation and embeddings are fundamental concepts in natural language processing (NLP) that involve transforming words or sentences into numerical vectors. These numerical representations enable computers to understand and process human language, making it easier to apply machine learning algorithms to NLP tasks. Let's explore language representation and embeddings in more detail.

Word2Vec and GloVe are both popular techniques used for word embedding, a process of representing words as dense vectors in a high-dimensional vector space. These word embeddings capture semantic relationships between words and are widely used in natural language processing tasks.

Word2Vec

Word2Vec is a family of word embedding models introduced by Mikolov et al. in 2013. It consists of two primary architectures: continuous bag of words (CBOW) and skip-gram:

1. CBOW: The CBOW model predicts a target word based on its context words. It takes a set of context words as input and tries to predict the target word in the middle of the context. It is efficient and can handle multiple context words in one shot.

2. Skip-gram: The skip-gram model does the opposite of CBOW. It takes a target word as input and tries to predict the context words around it. Skip-gram is useful for capturing word relationships and is known for performing better on rare words.

Word2Vec uses a shallow neural network with a single hidden layer to learn the word embeddings. The learned embeddings place semantically similar words closer together in the vector space.

GloVe (Global Vectors for Word Representation)

GloVe is another popular word embedding technique introduced by Pennington et al. in 2014. Unlike Word2Vec, GloVe uses a co-occurrence matrix of word pairs to learn word embeddings. The co-occurrence matrix represents how often two words appear together in a given corpus.

GloVe aims to factorize this co-occurrence matrix to obtain word embeddings that capture the global word-to-word relationships in the entire corpus. It leverages both global and local context information to create more meaningful word representations.

Now, let's resume the evolution of neural networks to LLMS in detail.

Probabilistic Models

The n-gram probabilistic model is a simple and widely used approach for language modeling in natural language processing (NLP). It estimates the probability of a word based on the preceding n-1 words in a sequence. The "n" in n-gram represents the number of words considered together as a unit. The n-gram model is built on the Markov assumption, which assumes that the probability of a word only depends on a fixed window of the previous words:

1. N-gram Representation: The input text is divided into contiguous sequences of n words. Each sequence of n words is treated as a unit or n-gram. For example, in a bigram model (n=2), each pair of consecutive words becomes an n-gram.

2. Frequency Counting: The model counts the occurrences of each n-gram in the training data. It keeps track of how often each specific sequence of words appears in the corpus.

3. Calculating Probabilities: To predict the probability of the next word in a sequence, the model uses the n-gram counts. For example, in a bigram model, the probability of a word is estimated based on the frequency of the preceding word (unigram). The probability is calculated as the ratio of the count of the bigram to the count of the unigram.

4. Smoothing: In practice, the n-gram model may encounter unseen n-grams (sequences not present in the training data). To handle this issue, smoothing techniques are applied to assign small probabilities to unseen n-grams.

5. Language Generation: Once the n-gram model is trained, it can be used for language generation. Starting with an initial word, the model predicts the next word based on the highest probabilities of the available n-grams. This process can be iteratively repeated to generate sentences.

The hidden Markov model (HMM) is another important probabilistic model in language processing. It is used to model data sequences that follow a Markovian structure, where an underlying sequence of hidden states generates observable events. The term "hidden" refers to the fact that we cannot directly observe the states, but we can infer them from the observable events. HMMs are used in various tasks, such as speech recognition, part-of-speech tagging, and machine translation.

Limitations:

– The n-gram model has limited context, considering only the preceding n-1 words, which may not capture long-range dependencies.

– It may not effectively capture semantic meaning or syntactic structures in the language.

Despite its simplicity and limitations, the n-gram probabilistic model provides a useful baseline for language modeling tasks and has been a foundational concept for more sophisticated language models like recurrent neural networks (RNNs) and Transformer-based models.

Neural Network–Based Language Models

Neural network–based language models have brought a significant breakthrough in natural language processing (NLP) in recent times. These models utilize neural networks, which are computational structures inspired by the human brain, to process and understand language.

The main idea behind these models is to train a neural network to predict the next word in a sentence based on the words that precede it. By presenting the network with a large amount of text data and teaching it to recognize patterns and relationships between words, it learns to make probabilistic predictions about what word is likely to come next.

Once the neural network is trained on a vast dataset, it can use the learned patterns to generate text, complete sentences, or even answer questions based on the context it has learned during training.

By effectively capturing the relationships and dependencies between words in a sentence, these language models have drastically improved the ability of computers to understand and generate human language, leading to significant advancements in various NLP applications like machine translation, sentiment analysis, chatbots, and much more.

Input Layer (n1, n2, ..., n_input)

＼＼　　＼

Hidden Layer (n3, n4, ..., n_hidden)

＼＼　　＼

Output Layer (n5, n6, ..., n_output)

In this diagram:

- "n_input" represents the number of input neurons, each corresponding to a feature in the input data.

- "n_hidden" represents the number of neurons in the hidden layer. The hidden layer can have multiple neurons, typically leading to more complex representations of the input data.

- "n_output" represents the number of neurons in the output layer. The number of output neurons depends on the nature of the problem—it could be binary (one neuron) or multiclass (multiple neurons).

Recurrent Neural Networks (RNNs)

Recurrent neural networks (RNNs) are a type of artificial neural network designed to process sequential data one element at a time while maintaining an internal state that summarizes the history of previous inputs. They have the unique ability to handle variable-length input and output sequences, making them well-suited for natural language processing tasks like language synthesis, machine translation, and speech recognition.

The key feature that sets RNNs apart is their capacity to capture temporal dependencies through feedback loops. These loops allow the network to use information from prior outputs as inputs for future predictions. This memory-like capability enables RNNs to retain context and information from earlier elements in the sequence, influencing the generation of subsequent outputs.

However, RNNs do face some challenges. The vanishing gradient problem is a significant issue, where the gradients used to update the network's weights become very small during training, making it difficult to learn long-term dependencies effectively. Conversely, the exploding gradient problem can occur when gradients become too large, leading to unstable weight updates.

Furthermore, RNNs are inherently sequential, processing elements one by one, which can be computationally expensive and challenging to parallelize. This limitation can hinder their scalability when dealing with large datasets.

To address some of these issues, more advanced variants of RNNs, such as long short-term memory (LSTM) and gated recurrent unit (GRU), have been developed. These variants have proven to be more effective at capturing long-term dependencies and mitigating the vanishing gradient problem.

RNNs are powerful models for handling sequential data, but they come with certain challenges related to long-term dependency learning, gradient issues, and computational efficiency. Their variants, like LSTM and GRU, have improved upon these limitations and remain essential tools for a wide range of sequential tasks in natural language processing and beyond.

Long Short-Term Memory (LSTM)

Long short-term memory (LSTM) networks are a specialized type of recurrent neural network (RNN) architecture designed to address the vanishing gradient problem and capture long-term dependencies in sequential data. They were introduced by Hochreiter and Schmidhuber in 1997 and have since gained popularity for modeling sequential data in various applications.

The key feature that sets LSTM apart from traditional RNNs is its ability to incorporate a memory cell that can selectively retain or forget information over time. This memory cell is controlled by three gates: the input gate, the forget gate, and the output gate:

- The input gate regulates the flow of new data into the memory cell, allowing it to decide which new information is important to store.

- The forget gate controls the retention of current data in the memory cell, allowing it to forget irrelevant or outdated information from previous time steps.

- The output gate regulates the flow of information from the memory cell to the network's output, ensuring that the relevant information is used in generating predictions.

This gating mechanism enables LSTM to capture long-range dependencies in sequential data, making it particularly effective for tasks involving natural language processing, such as language modeling, machine translation, and sentiment analysis. Additionally, LSTMs have been successfully applied in other tasks like voice recognition and image captioning.

By addressing the vanishing gradient problem and providing a better way to retain and utilize important information over time, LSTM networks have become a powerful tool for handling sequential data and have significantly improved the performance of various applications in the field of machine learning and artificial intelligence.

Gated Recurrent Unit (GRU)

GRU (gated recurrent unit) networks are a type of neural network architecture commonly used in deep learning and natural language processing (NLP). They are designed to address the vanishing gradient problem, just like LSTM networks.

Similar to LSTMs, GRUs also incorporate a gating mechanism, allowing the network to selectively update and forget information over time. This gating mechanism is crucial for capturing long-term dependencies in sequential data and makes GRUs effective for tasks involving language and sequential data.

The main advantage of GRUs over LSTMs lies in their simpler design and fewer parameters. This simplicity makes GRUs faster to train and more straightforward to deploy, making them a popular choice in various applications.

While both GRUs and LSTMs have a gating mechanism, the key difference lies in the number of gates used to regulate the flow of information. LSTMs use three gates: the input gate, the forget gate, and the output gate. In contrast, GRUs use only two gates: the reset gate and the update gate.

The reset gate controls which information to discard from the previous time step, while the update gate determines how much of the new information to add to the memory cell. These two gates allow GRUs to control the flow of information effectively without the complexity of having an output gate.

GRU networks are a valuable addition to the family of recurrent neural networks. Their simpler design and efficient training make them a practical choice for various sequence-related tasks, and they have proven to be highly effective in natural language processing, speech recognition, and other sequential data analysis applications.

Encoder-Decoder Networks

The encoder-decoder architecture is a type of neural network used for handling sequential tasks like language translation, chatbot, audio recognition, and image captioning. It is composed of two main components: the encoder network and the decoder network.

During language translation, for instance, the encoder network processes the input sentence in the source language. It goes through the sentence word by word, generating a fixed-length representation called the context vector. This context vector contains important information about the input sentence and serves as a condensed version of the original sentence.

Next, the context vector is fed into the decoder network. The decoder network utilizes the context vector along with its internal states to start generating the output sequence, which in this case is the translation in the target language. The decoder generates one word at a time, making use of the context vector and the previously generated words to predict the next word in the translation.

Sequence-to-Sequence Models

Sequence-to-sequence (Seq2Seq) models are a type of deep learning architecture designed to handle variable-length input sequences and generate variable-length output sequences. They have become popular in natural language processing (NLP) tasks like machine translation, text summarization, chatbots, and more. The architecture comprises an encoder and a decoder, both of which are recurrent neural networks (RNNs) or Transformer-based models.

Encoder

The encoder takes the input sequence and processes it word by word, producing a fixed-size representation (context vector) that encodes the entire input sequence. The context vector captures the essential information from the input sequence and serves as the initial hidden state for the decoder.

Decoder

The decoder takes the context vector as its initial hidden state and generates the output sequence word by word. At each step, it predicts the next word in the sequence based on the context vector and the previously generated words. The decoder is conditioned on the encoder's input, allowing it to produce meaningful outputs.

Attention Mechanism

In the standard encoder-decoder architecture, the process begins by encoding the input sequence into a fixed-length vector representation. This encoding step condenses all the information from the input sequence into a single fixed-size vector, commonly known as the "context vector."

The decoder then takes this context vector as input and generates the output sequence, step by step. The decoder uses the context vector and its internal states to predict each element of the output sequence.

While this approach works well for shorter input sequences, it can face challenges when dealing with long input sequences. The fixed-length encoding may lead to information loss because the context vector has a limited capacity to capture all the nuances and details present in longer sequences.

In essence, when the input sequences are long, the fixed-length encoding may struggle to retain all the relevant information, potentially resulting in a less accurate or incomplete output sequence.

To address this issue, more advanced techniques have been developed, such as using attention mechanisms in the encoder-decoder architecture. Attention mechanisms allow the model to focus on specific parts of the input sequence while generating each element of the output sequence. This way, the model can effectively handle long input sequences and avoid information loss, leading to improved performance and more accurate outputs.

The attention mechanism calculates attention scores between the decoder's hidden state (query) and each encoder's hidden state (key). These attention scores determine the importance of different parts of the input sequence, and the context vector is then formed as a weighted sum of the encoder's hidden states, with weights determined by the attention scores.

The Seq2Seq architecture, with or without attention, allows the model to handle variable-length sequences and generate meaningful output sequences, making it suitable for various NLP tasks that involve sequential data.

Training Sequence-to-Sequence Models

Seq2Seq models are trained using pairs of input sequences and their corresponding output sequences. During training, the encoder processes the input sequence, and the decoder generates the output sequence. The model is optimized to minimize the difference between the generated output and the ground truth output using techniques like teacher forcing or reinforcement learning.

Challenges of Sequence-to-Sequence Models

Seq2Seq models have some challenges, such as handling long sequences, dealing with out-of-vocabulary words, and maintaining context over long distances. Techniques like attention mechanisms and beam search have been introduced to address these issues and improve the performance of Seq2Seq models.

Sequence-to-sequence models are powerful deep learning architectures for handling sequential data in NLP tasks. Their ability to handle variable-length input and output sequences makes them well-suited for applications involving natural language understanding and generation.

Transformer

The Transformer architecture was introduced by Vaswani et al. in 2017 as a groundbreaking neural network design widely used in natural language processing tasks like text categorization, language modeling, and machine translation.

At its core, the Transformer architecture resembles an encoder-decoder model. The process begins with the encoder, which takes the input sequence and generates a hidden representation of it. This hidden representation contains essential information about the input sequence and serves as a contextualized representation.

The hidden representation is then passed to the decoder, which utilizes it to generate the output sequence. Both the encoder and decoder consist of multiple layers of self-attention and feed-forward neural networks.

The self-attention layer computes attention weights between all pairs of input components, allowing the model to focus on different parts of the input sequence as needed. The attention weights are used to compute a weighted sum of the input elements, providing the model with a way to selectively incorporate relevant information from the entire input sequence.

The feed-forward layer further processes the output of the self-attention layer with nonlinear transformations, enhancing the model's ability to capture complex patterns and relationships in the data.

The Transformer design offers several advantages over prior neural network architectures:

1. Efficiency: It enables parallel processing of the input sequence, making it faster and more computationally efficient compared to traditional sequential models.

2. Interpretability: The attention weights can be visualized, allowing us to see which parts of the input sequence the model focuses on during processing, making it easier to understand and interpret the model's behavior.

3. Global Context: The Transformer can consider the entire input sequence simultaneously, allowing it to capture long-range dependencies and improve performance on tasks like machine translation, where the context from the entire sentence is crucial.

The Transformer architecture has become a dominant approach in natural language processing and has significantly advanced the state of the art in various language-related tasks, thanks to its efficiency, interpretability, and ability to capture global context in the data.

Large Language Models (LLMs)

Large Language Models (LLMs) refer to a class of advanced artificial intelligence models specifically designed to process and understand human language at an extensive scale. These models are typically built using deep learning techniques, particularly Transformer-based architectures, and are trained on vast amounts of textual data from the Internet.

The key characteristic of large language models is their ability to learn complex patterns, semantic representations, and contextual relationships in natural language. They can generate humanlike text, translate between languages, answer questions, perform sentiment analysis, and accomplish a wide range of natural language processing tasks.

One of the most well-known examples of large language models is OpenAI's GPT (Generative Pre-trained Transformer) series, which includes models like GPT-3. These models are pre-trained on massive datasets and can be fine-tuned for specific applications, allowing them to adapt and excel in various language-related tasks.

The capabilities of large language models have brought significant advancements to natural language processing, making them instrumental in various industries, including customer support, content generation, language translation, and more. However, they also raise important concerns regarding ethics, bias, and misuse due to their potential to generate humanlike text and spread misinformation if not used responsibly.

Some notable examples of LLMs include the following:

1. GPT: GPT is the fourth version of OpenAI's Generative Pre-trained Transformer series. It is known for its ability to generate humanlike text and has demonstrated proficiency in answering questions, creating poetry, and even writing code.

2. BERT (Bidirectional Encoder Representations from Transformers): Developed by Google, BERT is a pivotal LLM that captures context from both directions of the input text, making it adept at understanding language nuances and relationships. It has become a foundational model for a wide range of NLP tasks.

3. T5 (Text-to-Text Transfer Transformer): Also developed by Google, T5 approaches all NLP tasks as text-to-text problems. This unifying framework has shown outstanding performance in tasks like translation, summarization, and question answering.

4. RoBERTa: Facebook's RoBERTa is an optimized version of BERT that has achieved state-of-the-art results across various NLP benchmarks. It builds upon BERT's architecture and training process, further improving language understanding capabilities.

These LLMs have demonstrated advancements in natural language processing, pushing the boundaries of what AI models can achieve in tasks like language generation, comprehension, and translation. Their versatility and state-of-the-art performance have made them valuable assets in applications ranging from chatbots and language translation to sentiment analysis and content generation. As research in the field progresses, we can expect even more sophisticated and capable LLMs to emerge, continuing to revolutionize the field of NLP.

Conclusion

The development of neural networks for large language models has brought about significant breakthroughs in the field of natural language processing (NLP).

From traditional probabilistic models like n-grams and hidden Markov models to more advanced neural network–based models such as recurrent neural networks (RNNs), long short-term memory (LSTM) networks, and gated recurrent units (GRUs), researchers have continuously improved these models to overcome challenges like vanishing gradients and handling large datasets efficiently.

One notable advancement is the introduction of attention-based techniques, particularly the Transformer architecture. Transformers have shown exceptional performance in various NLP applications by allowing the model to focus on specific parts of the input sequence using self-attention mechanisms.

These models have achieved remarkable success in language modeling because of their ability to effectively attend to different regions of the input sequence, capturing complex patterns and dependencies.

Lastly, the focus has shifted toward large language models (LLMs), which use deep neural networks to generate natural language text. LLMs like GPT-3 have demonstrated astonishing capabilities, generating humanlike text, answering questions, and performing various language-related tasks.

In conclusion, the advancements in neural networks for large language models have revolutionized the NLP landscape, enabling machines to understand and generate human language at an unprecedented level, opening up new possibilities for communication, content creation, and problem-solving.

In the coming chapters, let's deep dive into large language models architecture and applications.

CHAPTER 3

LLMs and Transformers

In this chapter, we embark on an enlightening journey into the world of LLMs and the intricacies of the Transformer architecture, unraveling the mysteries behind their extraordinary capabilities. These pioneering advancements have not only propelled the field of NLP to new heights but have also revolutionized how machines perceive, comprehend, and generate language.

The Power of Language Models

Language models have emerged as a driving force in the realm of natural language processing (NLP), wielding the power to transform how machines interpret and generate human language. These models act as virtual linguists, deciphering the intricacies of grammar, syntax, and semantics, to make sense of the vast complexities of human communication. The significance of language models lies not only in their ability to understand text but also in their potential to generate coherent and contextually relevant responses, blurring the lines between human and machine language comprehension.

At the core of language models is the concept of conditional probability, wherein a model learns the likelihood of a word or token occurring given the preceding words in a sequence. By training on extensive datasets containing a wide array of language patterns, these models become adept at predicting the most probable next word in a given context. This predictive power makes them indispensable in a myriad of NLP tasks, from machine translation and summarization to sentiment analysis, question answering, and beyond.

However, traditional language models had inherent limitations, especially when dealing with long-range dependencies and capturing the contextual nuances of language. The need for more sophisticated solutions paved the way for large language models (LLMs), which have revolutionized the field of NLP through their immense scale, powerful architectural innovations, and the remarkable abilities they possess.

© Akshay Kulkarni, Adarsha Shivananda, Anoosh Kulkarni, Dilip Gudivada 2023
A. Kulkarni et al., *Applied Generative AI for Beginners*, https://doi.org/10.1007/978-1-4842-9994-4_3

Large language models leverage massive computational resources and enormous amounts of data during their training process, enabling them to grasp the subtle intricacies of human language. Moreover, they excel at generalization, learning from the vast array of examples they encounter during pre-training and fine-tuning processes, which allows them to perform impressively on a wide range of NLP tasks.

The introduction of the Transformer architecture heralded a pivotal moment in the advancement of language models. Proposed in the seminal paper "Attention Is All You Need," the Transformer introduced the attention mechanism—a revolutionary concept that empowers the model to dynamically weigh the relevance of each word in a sequence concerning all other words. This attention mechanism, alongside feed-forward neural networks, forms the foundation of the Transformer's remarkable performance.

As language models continue to evolve, they hold the promise of driving even more profound advancements in AI-driven language understanding and generation. Nevertheless, with such power comes the responsibility to address ethical concerns surrounding biases, misinformation, and privacy. Striking a balance between pushing the boundaries of language modeling while upholding ethical considerations is crucial to ensuring the responsible deployment and impact of these powerful tools.

In the following sections, we delve deeper into the architectural intricacies of large language models and the Transformer, exploring how they operate, their real-world applications, the challenges they present, and the potential they hold for reshaping the future of NLP and artificial intelligence.

Transformer Architecture

As mentioned earlier, the Transformer architecture is a crucial component of many state-of-the-art natural language processing (NLP) models, including ChatGPT. It was introduced in the paper titled "Attention Is All You Need" by Vaswani et al. in 2017. The Transformer revolutionized NLP by providing an efficient way to process and generate language using self-attention mechanisms. Let's delve into an in-depth explanation of the core Transformer architecture.

Motivation for Transformer

The motivation for the Transformer architecture stemmed from the limitations and inefficiencies of traditional sequential models, such as recurrent neural networks (RNNs) and long short-term memory (LSTM) networks. These sequential models process language input one token at a time, which leads to several issues when dealing with long-range dependencies and parallelization.

The key motivations for developing the Transformer architecture were as follows:

- Long-Term Dependencies: Traditional sequential models like RNNs and LSTMs face difficulties in capturing long-range dependencies in language sequences. As the distance between relevant tokens increases, these models struggle to retain and propagate information over long distances.

- Inefficiency in Parallelization: RNNs process language input sequentially, making it challenging to parallelize computations across tokens. This limitation hampers their ability to leverage modern hardware with parallel processing capabilities, such as GPUs and TPUs, which are crucial for training large models efficiently.

- Gradient Vanishing and Exploding: RNNs suffer from the vanishing and exploding gradient problems during training. In long sequences, gradients may become very small or very large, leading to difficulties in learning and convergence.

- Reducing Computation Complexity: Traditional sequential models have quadratic computational complexity with respect to the sequence length, making them computationally expensive for processing long sequences.

The Transformer architecture, with its self-attention mechanism, addresses these limitations and offers several advantages.

Architecture

The Transformer architecture represented earlier in Figure 3-1 uses a combination of stacked self-attention and point-wise, fully connected layers in both the encoder and decoder, as depicted in the left and right halves of the figure, respectively.

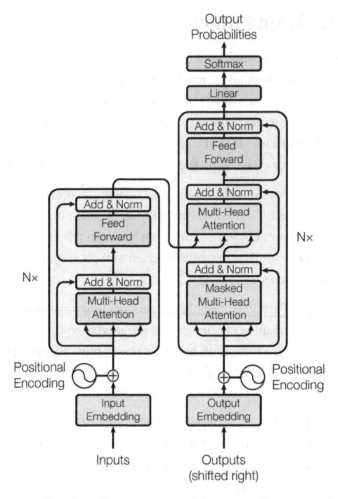

Figure 3-1. *The encoder-decoder structure of the Transformer architecture. Taken from "Attention Is All You Need" by Vaswani*

Encoder-Decoder Architecture

The Transformer architecture employs both the encoder stack and the decoder stack, each consisting of multiple layers, to process input sequences and generate output sequences effectively.

Encoder

The encoder represented earlier in Figure 3-2 is built with a stack of N = 6 identical layers, with each layer comprising two sub-layers. The first sub-layer employs a multi-head self-attention mechanism, allowing the model to attend to different parts of the input sequence simultaneously. The second sub-layer is a simple, position-wise fully connected feed-forward network, which further processes the output of the self-attention mechanism.

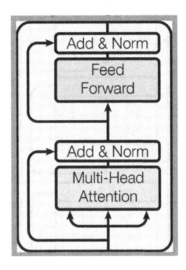

Figure 3-2. *The encoder-decoder structure of the Transformer architecture. Taken from "Attention Is All You Need"*

To ensure smooth information flow and facilitate learning, a residual connection is adopted around each of the two sub-layers. This means that the output of each sub-layer is added to the original input, allowing the model to learn and update the representations effectively.

To maintain the stability of the model during training, layer normalization is applied to the output of each sub-layer. This standardizes and normalizes the representations, preventing them from becoming too large or too small during the training process.

Furthermore, to enable the incorporation of residual connections, all sub-layers in the model, including the embedding layers, produce outputs of dimension dmodel = 512. This dimensionality helps in capturing the intricate patterns and dependencies within the data, contributing to the model's overall performance.

Decoder

The decoder shown earlier in Figure 3-3 in our model is structured similarly to the encoder, consisting of a stack of N = 6 identical layers. Each decoder layer, like the encoder layer, contains two sub-layers for multi-head self-attention and position-wise feed-forward networks. Conversely, the decoder introduces an additional third sub-layer, which utilizes multi-head attention to process the output of the encoder stack.

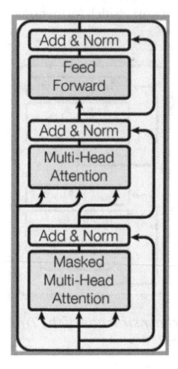

Figure 3-3. *The encoder-decoder structure of the Transformer architecture. Taken from "Attention Is All You Need"*

The purpose of this third sub-layer is to enable the decoder to access and leverage the contextualized representations generated by the encoder. By attending to the encoder's output, the decoder can align the input and output sequences, improving the quality of the generated output sequence.

To ensure effective learning and smooth information flow, the decoder, like the encoder, employs residual connections around each sub-layer, followed by layer normalization. This allows the model to maintain and propagate useful information effectively throughout the decoding process.

In contrast to the self-attention mechanism employed in the encoder, the self-attention sub-layer in the decoder is subject to a crucial modification. This alteration is designed to prevent positions within the sequence from attending to subsequent positions. The rationale behind this masking technique is pivotal in the realm of sequence-to-sequence tasks. Its primary objective is to ensure that the decoder generates output tokens in a manner known as "autoregression."

Autoregression is a fundamental concept in sequence generation tasks. It denotes that during the decoding process, the decoder is granted the capability to attend solely to the tokens it has previously generated. This deliberate restriction ensures that the decoder adheres to the correct sequential order when producing output tokens.

In practical terms, imagine the task of translating a sentence from one language to another. Autoregression guarantees that as the decoder generates each word of the translated sentence, it bases its decision on the words it has already translated. This mimics the natural progression of human language generation, where the context is built progressively, word by word. By attending only to prior tokens, the decoder ensures that it respects the semantic and syntactic structure of the output sequence, maintaining coherence and fidelity to the input.

In essence, autoregression is the mechanism that allows the decoder to "remember" what it has generated so far, ensuring that each subsequent token is contextually relevant and appropriately positioned within the sequence. It plays a pivotal role in the success of sequence-to-sequence tasks, where maintaining the correct order of token generation is of utmost importance.

To achieve this, the output embeddings of the decoder are offset by one position. As a result, the predictions for position "i" in the output sequence can only depend on the known outputs at positions less than "i." This mechanism ensures that the model generates the output tokens in an autoregressive manner, one token at a time, without access to information from future tokens.

By incorporating these modifications in the decoder stack, our model can effectively process and generate output sequences in sequence-to-sequence tasks, such as machine translation or text generation. The attention mechanism over the encoder's output empowers the decoder to align and contextually understand the input, while the autoregressive decoding mechanism guarantees the coherent generation of output tokens based on the learned context.

Attention

An attention function in the context of the Transformer architecture can be defined as a mapping between a query vector and a set of key–value pairs, resulting in an output vector. This function calculates the attention weights between the query and each key in the set and then uses these weights to compute a weighted sum of the corresponding values.

Here's a step-by-step explanation of the attention function:

Inputs

- Query Vector (Q): The query represents the element to which we want to attend. In the context of the Transformer, this is typically a word or token that the model is processing at a given time step.

- Key Vectors (K): The set of key vectors represents the elements that the query will attend to. In the Transformer, these are often the embeddings of the other words or tokens in the input sequence.

- Value Vectors (V): The set of value vectors contains the information associated with each key. In the Transformer, these are also the embeddings of the words or tokens in the input sequence.

Calculating Attention Scores

- The attention function calculates attention scores, which measure the relevance or similarity between the query and each key in the set.

- This is typically done by taking the dot product between the query vector (Q) and each key vector (K), capturing the similarity between the query and each key.

Calculating Attention Weights

- The attention scores are transformed into attention weights by applying the softmax function. The softmax function normalizes the scores, converting them into probabilities that sum up to 1.

- The attention weights represent the importance or relevance of each key concerning the query.

Weighted Sum

- The output vector is computed as the weighted sum of the value vectors (V), using the attention weights as the weights.

- Each value vector is multiplied by its corresponding attention weight, and all the weighted vectors are summed together to produce the final output vector.

- The output vector captures the contextual information from the value vectors based on the attention weights, representing the attended information relevant to the query.

The attention mechanism allows the model to selectively focus on the most relevant parts of the input sequence while processing each element (query). This ability to attend to relevant information from different parts of the sequence is a key factor in the Transformer's success in various natural language processing tasks as it enables the model to capture long-range dependencies and contextual relationships effectively.

Scaled Dot-Product Attention

The specific attention mechanism shown in Figure 3-4 employed in the Transformer is called "Scaled Dot-Product Attention," which is depicted in the preceding picture. Let's break down how Scaled Dot-Product Attention works:

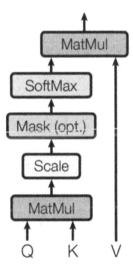

Figure 3-4. *The Scaled Dot-Product Attention structure of the Transformer architectureTaken from "Attention Is All You Need"*

Input and Matrices

- The input to Scaled Dot-Product Attention consists of queries (Q), keys (K), and values (V), each represented as vectors of dimension dk and dv.

- For each word in the input sequence, we create three vectors: a query vector, a key vector, and a value vector.

- These vectors are learned during the training process and represent the learned embeddings of the input tokens.

Dot Product and Scaling

- The Scaled Dot-Product Attention computes attention scores by performing the dot product between the query vector (Q) and each key vector (K).

- The dot product measures the similarity or relevance between the query and each key.

- The dot product of two vectors is the result of summing up the element-wise products of their corresponding components.

- To stabilize the learning process and prevent very large values in the dot product, the dot products are scaled down by dividing by the square root of the dimension of the key vector (`$\sqrt{dk}$`).

- This scaling factor of `$\sqrt{1/dk}$` is crucial in achieving stable and efficient attention computations.

Softmax and Attention Weights

- After calculating the scaled dot products, we apply the softmax function to transform them into attention weights.

- The softmax function normalizes the attention scores, converting them into probabilities that sum up to 1.

- The attention weights indicate the significance or relevance of each key in relation to the current query.

- Higher attention weights indicate that the corresponding value will contribute more to the final context vector.

Matrix Formulation and Efficiency

- Scaled Dot-Product Attention is designed for efficient computation using matrix operations.

- In practical applications, the attention function is performed on a set of queries (packed together into a matrix Q), keys (packed together into a matrix K), and values (packed together into a matrix V) simultaneously.

- The resulting matrix of outputs is then computed as follows:

$$\text{Attention}(Q, K, V) = \text{softmax}(QK^T / \sqrt{dk}) * V$$

  Where matrices Q are queries, K is keys, and V is values.

- This matrix formulation allows for highly optimized matrix multiplication operations, making the computation more efficient and scalable.

Scaled Dot-Product Attention has proven to be a critical component in the Transformer architecture, enabling the model to handle long-range dependencies and contextual information effectively. By attending to relevant information in the input sequence, the Transformer can create contextualized representations for each word, leading to remarkable performance in various natural language processing tasks, including machine translation, text generation, and language understanding. The use of matrix operations further enhances the computational efficiency of Scaled Dot-Product Attention, making the Transformer a powerful model for processing sequences of different lengths and complexities.

Multi-Head Attention

Multi-head attention shown earlier in Figure 3-5 is an extension of the Scaled Dot-Product Attention used in the Transformer architecture. It enhances the expressive power of the attention mechanism by applying multiple sets of attention computations in parallel, allowing the model to capture different types of dependencies and relationships in the input sequence.

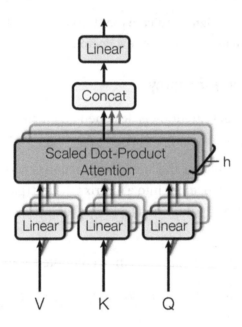

Figure 3-5. *The multi-head attention structure of the Transformer architectureTaken from "Attention Is All You Need"*

In the original Transformer paper ("Attention Is All You Need"), the authors introduced the concept of multi-head attention to overcome the limitations of single-headed attention, such as the restriction to a single attention pattern for all words. Multi-head attention allows the model to attend to different parts of the input simultaneously, enabling it to capture diverse patterns and dependencies.

Here's how multi-head attention works:

Input and Linear Projections

- Like in Scaled Dot-Product Attention, multi-head attention takes as input queries (Q), keys (K), and values (V), with each represented as vectors of dimension dk and dv.

- Instead of using the same learned projections for all attention heads, the input queries, keys, and values are linearly projected multiple times to create different sets of query, key, and value vectors for each attention head.

Multiple Attention Heads

- Multi-head attention introduces multiple attention heads, typically denoted by "h."

- Each attention head has its own set of linear projections to create distinct query, key, and value vectors.

- The number of attention heads, denoted as "h," is a hyperparameter and can be adjusted based on the complexity of the task and the model's capacity.

Scaled Dot-Product Attention per Head

- For each attention head, the Scaled Dot-Product Attention mechanism is applied independently, calculating attention scores, scaling, and computing attention weights as usual.

- This means that for each head, a separate context vector is derived using the attention weights.

Concatenation and Linear Projection

- After calculating the context vectors for each attention head, they are concatenated into a single matrix.

- The concatenated matrix is then linearly projected into the final output dimension.

Model's Flexibility

- By employing multiple attention heads, the model gains flexibility in capturing different dependencies and patterns in the input sequence.

- Each attention head can learn to focus on different aspects of the input, allowing the model to extract diverse and complementary information.

Multi-head attention is a powerful mechanism that enhances the expressive capacity of the Transformer architecture. It enables the model to handle various language patterns, dependencies, and relationships, leading to superior performance in complex natural language processing tasks. The combination of Scaled Dot-Product Attention with multiple attention heads has been a key factor in the Transformer's success and its ability to outperform previous state-of-the-art models in a wide range of NLP tasks.

The Transformer architecture utilizes multi-head attention in three distinct ways, each serving a specific purpose in the model's functioning:

1. Encoder-Decoder Attention:

   - In the encoder-decoder attention layers, the queries are generated from the previous decoder layer, representing the context from the current decoding step.

   - The memory keys and values are derived from the output of the encoder, representing the encoded input sequence.

   - This allows each position in the decoder to attend overall positions in the input sequence, enabling the model to align relevant information from the input to the output during the decoding process.

   - This attention mechanism mimics the typical encoder-decoder attention used in sequence-to-sequence models, which is fundamental in tasks like machine translation.

2. Encoder Self-Attention:

   - In the encoder, self-attention layers are applied, where all the keys, values, and queries are derived from the output of the previous layer in the encoder.

   - Each position in the encoder can attend to all positions in the previous layer of the encoder, allowing the model to capture dependencies and contextual relationships within the input sequence effectively.

   - Encoder self-attention is crucial for the model to understand the interdependencies of words in the input sequence.

3. Decoder Self-Attention with Masking:

   - The decoder also contains self-attention layers, but with a critical difference from encoder self-attention.

- In the decoder's self-attention mechanism, each position in the decoder can attend to all positions in the decoder up to and including that position.

- However, to preserve the autoregressive property (ensuring that each word is generated in the correct sequence), the model needs to prevent leftward information flow in the decoder.

- To achieve this, the input to the softmax function (which calculates attention weights) is masked by setting certain values to -∞ (negative infinity), effectively making some connections illegal.

- The masking prevents the model from attending to positions that would violate the autoregressive nature of the decoder, ensuring the generation of words in the correct order during text generation tasks.

Position-wise Feed-Forward Networks

Position-wise feed-forward networks (FFNs) are an essential component of the Transformer architecture, used in both the encoder and decoder layers. They play a key role in introducing nonlinearity and complexity to the model by processing each position in the input sequence independently and identically.

Example:

Given an input sequence X = {x_1, x_2, ..., x_seq_len} of shape (seq_len, d_model), where seq_len is the length of the sequence and d_model is the dimension of the word embeddings (e.g., d_model = 512):

1. Feed-Forward Architecture:

   The position-wise feed-forward network consists of two linear transformations with a ReLU activation function applied element-wise in between:

$$FFN\_1(X) = max(0, X * W1 + b1)$$

$$FFN\_Output = FFN\_1(X) * W2 + b2$$

Here, FFN_1 represents the output after the first linear transformation with weights W1 and biases b1. The ReLU activation function introduces nonlinearity by setting negative values to zero while leaving positive values unchanged. The final output FFN_Output is obtained after the second linear transformation with weights W2 and biases b2. This output is then element-wise added to the input as part of a residual connection.

2. Dimensionality:

The input and output of the position-wise feed-forward networks have a dimensionality of d_model = 512, which is consistent with the word embeddings in the Transformer model. The inner layer of the feed-forward network has a dimensionality of df f = 2048.

3. Parameter Sharing:

While the linear transformations are consistent across various positions in the sequence, each layer employs distinct learnable parameters. This design can also be thought of as two one-dimensional convolutions with a kernel size of 1.

Position-wise feed-forward networks enable the Transformer model to capture complex patterns and dependencies within the input sequence, complementing the attention mechanism. They introduce nonlinearity to the model, allowing it to learn and process information effectively, which has contributed to the Transformer's impressive performance in various natural language processing tasks.

Position Encoding

Positional encoding shown in Figure 3-6 is a critical component of the Transformer architecture, introduced to address the challenge of incorporating the positional information of words in a sequence. Unlike traditional recurrent neural networks (RNNs) that inherently capture the sequential order of words, Transformers operate on the entire input sequence simultaneously using self-attention. However, as self-attention does not inherently consider word order, positional encoding is necessary to provide the model with the positional information.

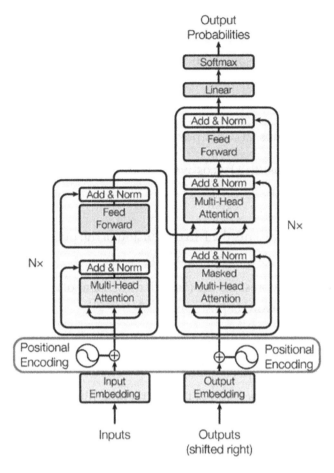

Figure 3-6. *The position encoding of the Transformer architectureTaken from "Attention Is All You Need"*

Importance of Positional Encoding:

- In the absence of positional encoding, the Transformer would treat the input as a "bag of words" without any notion of word order, which could result in the loss of sequential information.

- With positional encoding, the Transformer can distinguish between words in different positions, allowing the model to understand the relative and absolute positions of words within the sequence.

Formula for Positional Encoding:

The positional encoding is added directly to the input embeddings of the Transformer. It consists of sinusoidal functions of different frequencies to encode the position of each word in the sequence. The formula for the positional encoding is as follows:

$$PE(pos, 2i) = \sin(pos / 10000^{(2i/d\_model)})$$

$$PE(pos, 2i+1) = \cos(pos / 10000^{(2i/d\_model)})$$

Where

- "PE(pos, 2i)" represents the i-th dimension of the positional encoding for the word at position "pos."

- "PE(pos, 2i+1)" represents the (i+1)-th dimension of the positional encoding for the word at position "pos."

- "i" is the index of the dimension, ranging from 0 to "d_model - 1."

- The variable pos represents the position of the word in the sequence.

- "d_model" is the dimension of the word embeddings (e.g., d_model = 512).

Interpretation

The use of sine and cosine functions in the positional encoding introduces a cyclical pattern, allowing the model to learn different positional distances and generalizing to sequences of varying lengths. The positional encoding is added to the input embeddings before being passed through the encoder and decoder layers of the Transformer.

Positional encoding enriches the word embeddings with positional information, enabling the Transformer to capture the sequence's temporal relationships and effectively process the input data, making it one of the essential components that contributes to the Transformer's success in natural language processing tasks.

Advantages and Limitations of Transformer Architecture

Like any other architectural design, the Transformer has its advantages and limitations. Let's explore them:

Advantages

1. Parallelization and Efficiency: The Transformer's self-attention mechanism allows for parallel processing of input sequences, making it highly efficient and suitable for distributed computing, leading to faster training times compared to sequential models like RNNs.

2. Long-Range Dependencies: Thanks to the self-attention mechanism, the model can effectively capture long-range dependencies between words in a sequence.

3. Scalability: The Transformer's attention mechanism exhibits constant computational complexity with respect to the sequence length, making it more scalable than traditional sequential models, which often suffer from increasing computational costs for longer sequences.

4. Transfer Learning with Transformer: The Transformer architecture has demonstrated exceptional transferability in learning. Pre-trained models, such as BERT and GPT, serve as strong starting points for various natural language processing tasks. By fine-tuning these models on specific tasks, researchers and practitioners can achieve state-of-the-art results without significant architectural modifications. This transferability has led to widespread adoption and the rapid advancement of NLP applications.

5. Contextual Embeddings: The Transformer produces contextualized word embeddings, meaning that the meaning of a word can change based on its context in the sentence. This capability improves the model's ability to understand word semantics and word relationships.

6. Global Information Processing: Unlike RNNs, which process sequential information sequentially and may lose context over time, the Transformer processes the entire input sequence simultaneously, allowing for global information processing.

Limitations

1. Attention Overhead for Long Sequences: While the Transformer is efficient for parallelization, it still faces attention overhead for very long sequences. Processing extremely long sequences can consume significant computational resources and memory.

2. Lack of Sequential Order: The Transformer processes words in parallel, which might not fully exploit the inherent sequential nature of some tasks, leading to potential suboptimal performance for tasks where order matters greatly. Although positional encoding is used to provide positional information to the model, it does so differently from traditional RNNs. While it helps the Transformer understand the sequence's order, it does not capture it explicitly as RNNs do. This distinction is important to note in understanding how Transformers handle sequential information.

3. Excessive Parameterization: The Transformer has a large number of parameters, especially in deep models, which can make training more challenging, especially with limited data and computational resources.

4. Inability to Handle Unstructured Inputs: The Transformer is designed primarily for sequences, such as natural language sentences. It may not be the best choice for unstructured inputs like images or tabular data.

5. Fixed Input Length: For the most part, the Transformer architecture requires fixed-length input sequences due to the use of positional encodings. Handling variable-length sequences may require additional preprocessing or padding. It's worth noting that there are some length-adaptive variants of the Transformer architecture that offer more flexibility in this regard.

Conclusion

In conclusion, large language models (LLMs) based on the Transformer architecture have emerged as a groundbreaking advancement in the realm of natural language processing. Their ability to capture long-range dependencies, combined with extensive pre-training on vast datasets, has revolutionized natural language understanding tasks. LLMs have demonstrated remarkable performance across various language-related challenges, outperforming traditional approaches and setting new benchmarks. Moreover, they exhibit great potential in language generation and creativity, capable of producing humanlike text and engaging stories. However, alongside their numerous advantages, ethical considerations loom large, including concerns regarding biases, misinformation, and potential misuse. Researchers and engineers are actively working on addressing these challenges to ensure responsible AI deployment. Looking ahead, the future of LLMs and Transformers promises exciting opportunities, with potential applications in diverse domains like education, healthcare, customer support, and content generation. As the field continues to evolve, LLMs are poised to reshape how we interact with and comprehend language, opening new possibilities for transformative impact in the years to come.

## CHAPTER 4

The ChatGPT Architecture: An In-Depth Exploration of OpenAI's Conversational Language Model

In recent years, significant advancements in natural language processing (NLP) have paved the way for more interactive and humanlike conversational agents. Among these groundbreaking developments is ChatGPT, an advanced language model created by OpenAI. ChatGPT is based on the GPT (Generative Pre-trained Transformer) architecture and is designed to engage in dynamic and contextually relevant conversations with users.

ChatGPT represents a paradigm shift in the world of conversational AI, allowing users to interact with a language model in a more conversational manner. Its ability to understand context, generate coherent responses, and maintain the flow of conversation has captivated both researchers and users alike. As the latest iteration of NLP models, ChatGPT has the potential to transform how we interact with technology and information.

This chapter explores the intricacies of the ChatGPT architecture, delving into its underlying mechanisms, training process, and capabilities. We will uncover how ChatGPT harnesses the power of transformers, self-attention, and vast amounts of pre-training data to become an adept conversationalist. Additionally, we will discuss the strengths and limitations of ChatGPT, along with the ethical considerations surrounding its use. With ChatGPT at the forefront of conversational AI, this chapter aims to shed light on the fascinating world of state-of-the-art language models and their impact on the future of human–computer interaction.

© Akshay Kulkarni, Adarsha Shivananda, Anoosh Kulkarni, Dilip Gudivada 2023
A. Kulkarni et al., *Applied Generative AI for Beginners*, https://doi.org/10.1007/978-1-4842-9994-4_4

The Evolution of GPT Models

The evolution of the GPT (Generative Pre-trained Transformer) models has been marked by a series of significant advancements. Each new version of the model has typically featured an increase in the number of parameters and has been trained on a more diverse and comprehensive dataset. Here is a brief history:

1. GPT-1: The original GPT model, introduced by OpenAI in 2018, was based on the Transformer model. This model was composed of 12 layers, each with 12 self-attention heads and a total of 117 million parameters. It used unsupervised learning and was trained on the BookCorpus dataset, a collection of 7,000 unpublished books.

2. GPT-2: OpenAI released GPT-2 in 2019, which marked a significant increase in the scale of the model. It was composed of 48 layers and a total of 1.5 billion parameters. This version was trained on a larger corpus of text data scraped from the Internet, covering a more diverse range of topics and styles. However, due to concerns about potential misuse, OpenAI initially decided not to release the full model, instead releasing smaller versions and later releasing the full model as those concerns were addressed.

3. GPT-3: GPT-3, introduced in 2020, marked another significant step up in scale, with 175 billion parameters and multiple transformer layers. This model demonstrated an impressive ability to generate text that closely resembled human language. The release of GPT-3 spurred widespread interest in the potential applications of large language models, as well as discussions about the ethical implications and challenges of such powerful models.

4. GPT-4: GPT-4 is a revolutionary multimodal language model with capabilities extending to processing both text and image inputs, describing humor in images, and summarizing text from screenshots. GPT-4's interactions with external interfaces enable tasks beyond text prediction, making it a transformative tool in natural language processing and various domains.

Throughout this evolution, one of the key themes has been the power of scale: generally speaking, larger models trained on more data tend to perform better. However, there's also been increasing recognition of the challenges associated with larger models, such as the potential for harmful outputs, the increased computational resources required for training, and the need for robust methods for controlling the behavior of these models.

The Transformer Architecture: A Recap

As mentioned earlier in the previous chapter, we have already explored the Transformer architecture shown in Figure 4-1 in detail. This concise summary serves as a recap of the key components for those readers who are already familiar with the Transformer architecture. For a more comprehensive understanding, readers can refer back to the earlier chapter where the Transformer architecture was thoroughly explained with its components and working mechanisms.

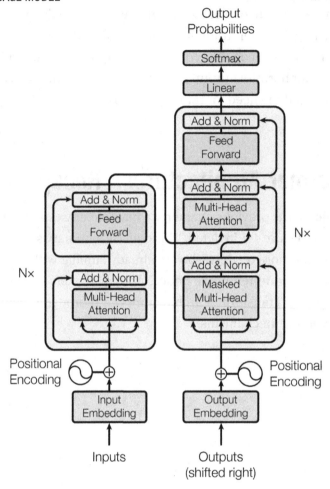

Figure 4-1. *The encoder-decoder structure of the Transformer architectureTaken from "Attention Is All You Need"*

Here are some key pointers to remember about the Transformer architecture:

- The Transformer architecture revolutionized natural language processing with its attention-based mechanism.

- Key components of the Transformer include the self-attention mechanism, encoder-decoder structure, positional encoding, multi-head self-attention, and feed-forward neural networks.

- Self-attention allows the model to weigh the importance of different words and capture long-range dependencies.

- The encoder-decoder structure is commonly used in machine translation tasks.

- Positional encoding is used to incorporate word order information into the input sequence.

- Multi-head self-attention allows the model to attend to multiple parts of the input simultaneously, enhancing its ability to capture complex relationships within the data.

- Feed-forward neural networks process information from the attention layers.

- Residual connections and layer normalization stabilize training in deep architectures.

Architecture of ChatGPT

The GPT architecture plays a foundational role in enabling the capabilities of ChatGPT as an interactive conversational AI. While we have already explored the Transformer architecture in the previous chapter, this section delves into how it is specifically adapted and optimized for chat-based interactions in ChatGPT. ChatGPT, like all models in the GPT series, is based on a Transformer architecture, specifically leveraging a "decoder-only" structure from the original Transformer model. Additionally, ChatGPT incorporates a crucial component known as "reinforcement learning from human feedback (RLHF)." RLHF is an advanced technique that enhances ChatGPT's performance, and it will be covered in detail later in this chapter, providing you with a comprehensive understanding of its significance, as shown in Figure 4-2.

Figure 4-2 presents an architecture diagram of ChatGPT, illustrating its training process in detail. This diagram provides a comprehensive view of how ChatGPT learns and refines its capabilities during the training phase. It showcases the flow of data, the model's internal components, and the training pipeline, offering insights into the model's development.

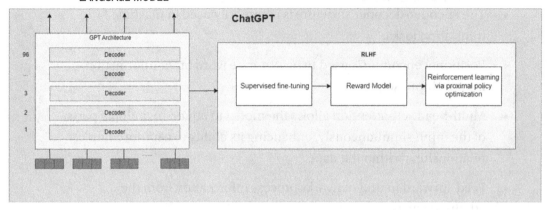

Figure 4-2. *ChatGPT architecture*

Here's an overview of the key elements:

1.   Transformer Models:

     Transformer models are a type of model used in machine learning,
     particularly in the field of natural language processing (NLP). They
     were introduced by Vaswani et al. in the paper "Attention is All You
     Need." The main advantage of Transformer models is that they
     process input data in parallel rather than sequentially, allowing
     for more efficient computation and the ability to handle longer
     sequences of data. They also introduced the concept of "attention,"
     enabling the model to weigh the importance of different words in
     the input when generating an output.

2.   Decoder-Only Structure:

     The initial Transformer model presented by Vaswani et al.
     included two parts: an encoder, which processes the input, and a
     decoder, which generates the output. However, GPT models like
     ChatGPT use only the decoder part shown in Figure 4-3 of the
     Transformer architecture.

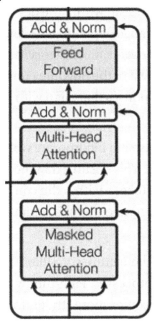

Figure 4-3. *The decoder structure of the Transformer architectureTaken from
"Attention Is All You Need"*

This results in a unidirectional structure, where each token (or
word) can only attend to earlier positions in the input sequence.
This design allows GPT models to generate text one word at
a time, using the words it has already generated to inform the
generation of the next word. This design choice is driven by the
nature of the conversational AI task, where the model needs to
generate responses based on the input conversation history.

The decoder layer in ChatGPT is responsible for generating the
next token in the response sequence given the context of the
conversation history. It employs a combination of self-attention
and feed-forward neural networks to process the input tokens and
generate meaningful and contextually relevant replies.

The self-attention mechanism within the decoder allows the model to capture long-range dependencies and relationships between tokens in the conversation history. This is critical for understanding the context of the ongoing conversation and producing coherent responses that align with the preceding dialogue.

Positional encoding is used to incorporate word order information into the input sequence. This ensures that the model understands the relative positions of tokens within the conversation history, enabling it to generate responses that are contextually appropriate.

Using a decoder-only architecture simplifies the model's training and inference processes. Fine-tuning the decoder for conversational tasks becomes more straightforward, as the focus is solely on generating responses based on the provided context.

Additionally, the decoder-only setup in ChatGPT makes it more efficient for real-time interactions. By eliminating the encoder, computational resources are focused solely on the decoder, allowing for quicker response times during conversations.

Furthermore, ChatGPT leverages techniques like reinforcement learning from human feedback to optimize the decoder's performance. Fine-tuning the model with human-generated responses and feedback aligns the model's outputs with desired human preferences, improving the quality of the generated responses.

Overall, the decision to use a decoder-only architecture in ChatGPT is a carefully considered technical choice, tailored to the conversational AI context. It enables the model to generate accurate and contextually appropriate responses efficiently, making it a powerful tool for interactive and engaging chat-based applications.

3. Self-Attention Mechanism:

The self-attention mechanism is a key element of the Transformer architecture. In self-attention, each token in the input can interact with every other token, rather than only adjacent or nearby tokens. This allows the model to better capture the context of each word in a sentence. In ChatGPT, the self-attention mechanism is utilized within the decoder layers to capture dependencies and relationships between tokens in the conversation history, enabling the model to understand the context and generate relevant responses.

Here's how the self-attention mechanism works in ChatGPT:

- Contextual Understanding: In a conversation, each word or token depends on other words within the conversation history to gain its contextual meaning. The self-attention mechanism allows the model to pay attention to all the tokens in the conversation history and weigh their importance in generating the next token. This helps the model to understand the ongoing context and produce responses that are coherent and contextually relevant.

- Attention Scores: During self-attention, the model computes attention scores that indicate the importance of each token with respect to the current token being processed. Tokens that are more relevant in the context of the current token receive higher attention scores, while less relevant tokens receive lower scores. This dynamic weighting of tokens allows the model to focus on the most relevant parts of the conversation history for generating the response.

- Capturing Long-Range Dependencies: The self-attention mechanism enables ChatGPT to capture long-range dependencies in the conversation history. Unlike traditional recurrent neural networks, which have limited memory, the self-attention mechanism allows the model to consider all tokens in the conversation history regardless of their distance from the current token. This capability is crucial for understanding the flow of the conversation and generating responses that maintain coherence over extended dialogues.

- Positional Encoding: In the Transformer architecture, including ChatGPT, positional encoding is introduced to incorporate the order of tokens into the self-attention process. Positional encoding ensures that the model understands the sequential order of tokens within the conversation history, allowing it to differentiate between different positions in the dialogue and make contextually appropriate predictions.

4. Layered Structure: ChatGPT's architecture consists of multiple layers of these Transformer decoders stacked on top of each other. Each layer learns to represent the input data in a way that helps the subsequent layer to better perform the task. The number of layers can vary across different versions of GPT; for instance, GPT-3 has 96 Transformer layers.

Here's how the layered structure works in ChatGPT:

- Stacked Decoder Layers: ChatGPT employs a decoder-only architecture, meaning that only the decoder layers are used and the encoder layers are omitted. The conversation history serves as the input to the decoder, and the model's objective is to generate the next token in the response sequence based on this input context. The decoder layers are stacked on top of each other, and the number of layers can vary based on the model's configuration.

- Hierarchical Feature Extraction: Each decoder layer in ChatGPT performs a series of operations on the input tokens. The self-attention mechanism in each layer allows the model to attend to all tokens in the conversation history, capturing relevant information and dependencies across the entire sequence. This hierarchical feature extraction enables the model to progressively refine its understanding of the context as it moves through the layers.

- Positional Encoding: To handle the sequential nature of the input data, positional encoding is incorporated into each layer. This encoding provides information about the order and position of tokens within the conversation history, ensuring that the model can differentiate between tokens and understand their positions in the dialogue.

- Feed-Forward Neural Networks: After the self-attention step, the model further processes the tokens using feed-forward neural networks within each layer. These networks apply linear transformations and nonlinear activations to the tokens, enabling the model to capture complex patterns and relationships within the conversation.

- Residual Connections and Layer Normalization: Residual connections and layer normalization are used in each decoder layer to stabilize the training process and facilitate information flow. Residual connections, sometimes referred to as skip connections, allow the model to retain important information from the previous layers and provide a mechanism to "skip" some layers by zeroing the weights, resulting in an overspecified model that can learn sparsity. Layer normalization complements this by normalizing the inputs and outputs of each layer, contributing to improved training convergence.

By stacking multiple decoder layers, ChatGPT can capture increasingly complex patterns and contextual dependencies in the conversation history. This layered structure is crucial for the model's ability to generate coherent, contextually appropriate responses in chat-based interactions. The hierarchical feature extraction and the progressive refinement of information enable ChatGPT to perform effectively in a wide range of natural language processing tasks, making it a powerful conversational AI tool.

5. Positional Encodings: Since Transformer models process all input
    tokens in parallel, they do not inherently capture the sequential
    order of the data. To account for this, GPT models use positional
    encodings, which provide information about the position of each
    word in the sequence. This allows the model to understand the
    order of words and make accurate predictions based on that
    order. Therefore, while positional encodings are essential for the
    functioning of ChatGPT and other Transformer models, they
    are not unique to ChatGPT and are a fundamental part of the
    Transformer architecture itself.

6. Masked Self-Attention: In the decoder, the self-attention
    mechanism is modified to prevent tokens from attending to future
    tokens in the input sequence. This is known as "masked" self-
    attention. Masked self-attention is a crucial component of the
    Transformer architecture and is also used in ChatGPT to handle
    sequential data efficiently. In the context of ChatGPT, masked
    self-attention allows the model to attend to only the relevant
    tokens within the input sequence while preventing information
    flow from future positions. This is particularly important during
    autoregressive text generation to maintain causality and ensure
    that the model generates text sequentially, one token at a time.

    •   Masked Self-Attention in ChatGPT: In the transformer decoder
        layers of ChatGPT, each token attends to all other tokens in the
        input sequence, including itself, using self-attention. However,
        to prevent information from leaking from future tokens during
        generation, a masking mechanism is applied to the self-
        attention matrix.

    •   Masking Mechanism: The masking mechanism involves applying
        a triangular mask to the self-attention matrix, where all elements
        below the main diagonal are set to negative infinity (or a very
        large negative value). This effectively masks out the future
        tokens and allows the token to only attend to its previous tokens
        and itself.

- Example: Let's consider an example of generating the sentence "I love natural language processing" using ChatGPT. During the generation process, when the model is predicting the word "language," it should only attend to the previous tokens "I," "love," "natural," and "language" itself. The attention to the word "processing" should be masked out to maintain causality.

- Benefit in Autoregressive Text Generation: Masked self-attention ensures that ChatGPT generates text in an autoregressive manner, where each token's prediction only depends on previously generated tokens. This is crucial for generating coherent and grammatically correct sentences. Without masking, the model might have access to information from future tokens, leading to incorrect and nonsensical output.

7. Reinforcement Learning from Human Feedback (RLHF)

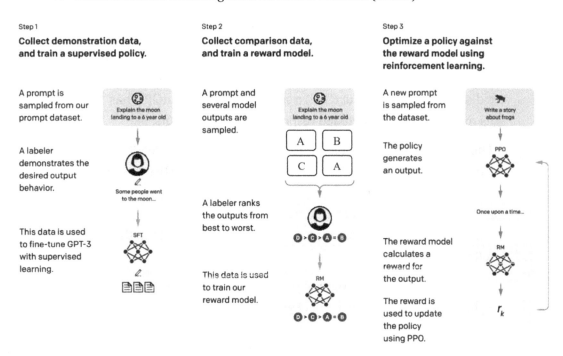

Figure 4-4. *Taken from "training language models to follow instructions with human feedback" where A (explain gravity), B (explain war), or C (moon) is a natural satellite of D (people went to the moon)*

Reinforcement learning from human feedback (RLHF) shown earlier in Figure 4-4 is a pivotal component of the ChatGPT architecture, playing a crucial role in its fine-tuning process and elevating its conversational capabilities. The RLHF approach enables ChatGPT to learn from human evaluators and adapt its language generation based on their feedback. RL, or reinforcement learning, is a type of machine learning where an agent learns by interacting with its environment and receiving feedback in the form of rewards. Unlike unsupervised learning, where the model learns from unlabeled data without any specific guidance, and supervised learning, where it's trained on labeled data with predefined correct answers, RL involves trial-and-error learning:

1. Supervised Fine-Tuning: Supervised fine-tuning is an essential phase in the development of ChatGPT. Initially, ChatGPT undergoes supervised fine-tuning, wherein human AI trainers simulate conversations by playing both the user and the AI assistant roles. During this process, trainers have access to model-written suggestions to help them generate responses that align with the desired conversational outcomes.

    This dialogue dataset, derived from supervised fine-tuning, is then combined with the InstructGPT dataset, which is transformed into a dialogue format. InstructGPT, a sibling model of ChatGPT, has its roots in providing detailed responses to user prompts.

    The connection to reinforcement learning from human feedback (RLHF) becomes apparent when we consider that RLHF takes this initial supervised training a step further. RLHF enables ChatGPT to learn and adapt through interactions with human evaluators who provide feedback, creating a continuous feedback loop that refines the model's responses over time.

    By understanding this progression from supervised fine-tuning, influenced by InstructGPT's background, to RLHF, we gain insight into how ChatGPT evolves and aligns its capabilities with human expectations in the realm of natural language understanding and generation.

2. Reward Model: The model trained via supervised learning
    is then used to collect comparison data. AI trainers engage
    in conversations with the chatbot and rank different model-
    generated responses by quality. This dataset is used as a reward
    model to guide the reinforcement learning process.

3. Reinforcement Learning via Proximal Policy Optimization:
    Reinforcement learning via proximal policy optimization is
    a crucial step in ChatGPT's development. In RL, a "policy"
    refers to a set of rules or strategies that an AI agent follows to
    make decisions in an environment. In this case, the chatbot,
    ChatGPT, has a "policy" that guides how it generates responses in
    conversations.

    During this phase, the model uses comparison data to improve its
    policy through a method called proximal policy optimization (PPO).
    PPO is a technique that optimizes the chatbot's policy with the goal of
    increasing the likelihood of generating better-rated responses while
    decreasing the likelihood of generating worse-rated ones.

    To connect this to the broader context, let's backtrack a bit.
    ChatGPT starts as a pre-trained model, which means it has a basic
    understanding of language from its initial training. However,
    to make it truly conversational and responsive, it undergoes
    a process of fine-tuning, where it refines its abilities based on
    human feedback.

    The reinforcement learning phase with PPO is a part of this
    fine-tuning process. It's like teaching the chatbot specific
    conversational strategies to ensure it provides high-quality
    responses. So, in essence, the connection here is that this
    reinforcement learning step further refines ChatGPT's "policy" to
    make it better at generating natural and engaging conversations.

    The model continues to iterate on this process, learning from
    the comparison data and using PPO to improve the responses
    it generates. This cycle is repeated, enabling the model to
    continuously improve its understanding and response quality
    based on human feedback.

In this way, RLHF plays a pivotal role in shaping ChatGPT's performance. It allows OpenAI to systematically improve the model based on direct human feedback, helping the model avoid incorrect responses, and better align its responses with human values.

This combination of supervised learning with RLHF provides a robust framework for training ChatGPT and similar models, blending the strengths of traditional machine learning with the nuanced feedback that only humans can provide.

To summarize, ChatGPT leverages the Transformer architecture, specifically a "decoder-only" structure and RLHF to efficiently process and generate text. The use of self-attention allows it to consider the full context of the input, while positional encodings ensure that the sequential order of words is captured. These aspects combine to allow ChatGPT to generate impressively humanlike text.

Pre-training and Fine-Tuning in ChatGPT

In the development of ChatGPT, two crucial stages play a pivotal role in shaping its capabilities: pre-training and fine-tuning. Pre-training involves language modeling on massive datasets to impart foundational language understanding to the model, while fine-tuning adapts the pre-trained model to specific tasks and user interactions, making it contextually relevant and effective in real-world scenarios.

Pre-training: Learning Language Patterns

The pre-training phase is the initial step in creating ChatGPT. During this stage, the model undergoes unsupervised learning on extensive and diverse datasets containing a wide range of text from various sources. Using the Transformer architecture, ChatGPT learns to predict the next word in a sequence based on the context of preceding words. By absorbing large amounts of text data, the model internalizes grammar, syntax, semantics, and contextual relationships, enabling it to generate coherent and contextually appropriate responses during interactions.

Fine-Tuning: Adapting to Specific Tasks

While pre-training equips ChatGPT with a broad understanding of language, it is not directly tailored to specific tasks or user interactions. The fine-tuning phase bridges this gap by adapting the pre-trained model to particular domains and tasks. During fine-tuning, ChatGPT is exposed to domain-specific datasets, which can include labeled examples for supervised learning or demonstrations of desired behavior:

- Domain Adaptation: Fine-tuning allows ChatGPT to adapt its knowledge to the domain it will be utilized in. For example, if ChatGPT is intended to assist with customer support, fine-tuning may involve exposure to customer service conversations and queries.

- User Interaction Guidance: In addition to domain adaptation, fine-tuning incorporates user interaction guidance to ensure ChatGPT responds contextually and responsibly to user inputs. This may involve reinforcement learning from human feedback to reinforce desired behaviors and discourage harmful or inappropriate responses.

Continuous Learning and Iterative Improvement

Pre-training and fine-tuning are not isolated events but part of an ongoing process of continuous learning and improvement. As ChatGPT interacts with users and receives feedback, it can further fine-tune its responses to specific user preferences and evolving context, enhancing its overall performance and responsiveness.

Contextual Embeddings in ChatGPT

Contextual embeddings form the foundation of language models like ChatGPT. Unlike traditional word embeddings such as Word2Vec or GloVe, which assign a fixed vector to each word regardless of its context, contextual embeddings provide a unique vector for each word based on its position and surrounding words in a sentence.

For ChatGPT, the contextual embedding of a word is computed from the self-attention mechanism of the transformer model. Given a sequence of words as input, the self-attention mechanism computes a weighted sum of the input word embeddings, where the weights are determined by the similarity between the current word and the other words in the sentence. This produces a unique embedding for each word that captures its specific role within the sentence.

The self-attention mechanism is applied in multiple layers, allowing the model to develop increasingly abstract representations of the input. The outputs of the final layer provide the contextual embeddings used to generate the next word in the sequence. Each word's contextual embedding incorporates information from all the previous words in the sentence, which allows the model to generate coherent and contextually appropriate responses.

Response Generation in ChatGPT

Once the contextual embeddings are computed, ChatGPT utilizes a process known as autoregressive generation to craft responses that are contextually appropriate and coherent. This process unfolds as follows.

Starting with a specialized start-of-sequence token, the model initiates the generation sequence. It predicts the next word in the sequence one word at a time, utilizing the previous words as context.

At each step, the model calculates a probability distribution over the entire vocabulary for the next word, grounded in the current contextual embedding. The choice of the next word can take several forms: it can be the word with the highest probability, known as "greedy decoding," introducing determinism; alternatively, it can be sampled from the distribution, introducing an element of unpredictability through "random sampling." Furthermore, ChatGPT can balance these approaches, employing techniques such as "top-k sampling" or "nucleus sampling," which select from the top-k highest probability words or a set of words with cumulative probabilities surpassing a certain threshold, respectively.

Once a word is selected, it is incorporated into the response sequence, and the contextual embeddings are promptly updated to encompass this newly chosen word. This process repeats iteratively, generating each subsequent word. It continues until ChatGPT generates an end-of-sequence token or reaches a predetermined maximum sequence length.

Crucially, this intricate response generation process unfolds within a unified ChatGPT architecture, dispelling any notion of separateness. The term "policy" within ChatGPT, which guides word selection and response construction, is not an isolated entity; rather, it consists of learned weights and parameters inherent to the model. These weights represent the model's understanding of language patterns, context, and suitable behavior, all gleaned during training. Therefore, when discussing the methods for word selection, it is an exploration of how these learned weights influence ChatGPT's behavior within a single integrated framework.

In essence, ChatGPT's response generation leverages this unified architecture and its policy to predict and generate words, culminating in responses that demonstrate both contextual coherence and relevance. It's important to clarify that the model's response generation is not driven by explicit understanding or planning; rather, it relies on its learned knowledge of statistical language patterns, all encapsulated within its policy.

Handling Biases and Ethical Considerations

Addressing Biases in Language Models

Language models like ChatGPT learn from large datasets that contain text from the Internet. Given the nature of these datasets, the models might pick up and propagate the biases present in the training data. These biases can manifest in various forms such as gender bias, racial bias, or bias toward controversial or sensitive topics. The biases could impact the way the AI system interacts with users, often leading to outputs that may be offensive, inappropriate, or politically biased.

Recognizing the potential harm these biases can cause is crucial. If unchecked, they can perpetuate harmful stereotypes, misinform users, and potentially alienate certain user groups.

OpenAI's Efforts to Mitigate Biases

OpenAI is fully aware of the potential for biases in AI system outputs and has been making concerted efforts to address them.

- Fine-Tuning with Human Supervision: After the initial pre-training, OpenAI uses a process of fine-tuning with human reviewers, who follow guidelines provided by OpenAI. The guidelines explicitly state not to favor any political group. The human reviewers review and rate possible model outputs for a range of example inputs. Through an iterative process, the model generalizes from reviewer feedback

to respond to a wide array of user inputs. However, this fine-tuning
process is resource-intensive, impacting both cost and the timeline
for AI model deployment.

- Regular Updates to Guidelines: The guidelines for human reviewers
  are not static and are updated regularly based on ongoing feedback
  from users and developments in society at large. OpenAI maintains
  a strong feedback loop with reviewers through weekly meetings to
  address questions and provide clarifications, which helps in training
  the model more effectively and reducing biases in its responses. Yet,
  achieving consensus on guidelines can be challenging in a constantly
  evolving linguistic landscape.

- Transparency: OpenAI is committed to being transparent about
  its intentions, progress, and the limitations of its models. The
  organization publishes regular updates and encourages public input
  on its technology, policies, and disclosure mechanisms. However,
  transparency has its limits due to the intricacies of AI systems and the
  necessity of safeguarding user privacy.

- Research and Development: OpenAI is currently conducting
  extensive research to minimize both overt and subtle biases in
  how ChatGPT generates responses to various inputs. This includes
  improvements in the clarity of guidelines regarding potential pitfalls
  and challenges tied to bias, as well as controversial figures and
  themes. These research initiatives aim to enhance AI's understanding
  of complex societal nuances.

- Customization and User Feedback: OpenAI is developing an
  upgrade to ChatGPT that allows users to easily customize their
  behavior, within broad societal limits. This way, AI can be a useful
  tool for individual users, without imposing a one-size-fits-all model.
  User feedback is actively encouraged and is invaluable in making
  necessary adjustments and improvements. However, customization
  introduces challenges related to defining these bounds of acceptable
  behavior and ensuring responsible AI usage.

However, it's evident that addressing biases in AI is not a straightforward task but rather a nuanced and intricate endeavor. OpenAI's approach involves fine-tuning with human supervision, regular updates to guidelines, transparency, research and development, and the introduction of customization options.

However, it's crucial to acknowledge that the pursuit of bias-free AI responses comes with trade-offs. These include increased costs, potential performance implications, and the challenge of aligning AI systems with ever-evolving language nuances. Additionally, the fundamental challenge of defining and achieving unbiased datasets and processes persists in this dynamic landscape.

OpenAI remains committed to continuous learning and improvement in the realm of bias mitigation. The organization recognizes that while these efforts help mitigate biases, they may not entirely eliminate them. As we move forward, it's important to engage in collaborative discussions, share feedback, and collectively work toward building AI systems that respect diverse perspectives and values.

Strengths and Limitations
Strengths of ChatGPT

- Understanding of Context: ChatGPT, with its Transformer-based architecture, has a strong understanding of context and can maintain the context of a conversation over several turns. It can generate humanlike text based on the context it has been provided with, making it a powerful tool for a range of applications, from drafting emails to creating written content, and even coding help.

- Large-Scale Language Model: As a large-scale language model, ChatGPT has been trained on diverse Internet text. Therefore, it has a broad knowledge base and can generate responses on a wide range of topics.

- Fine-Tuning Process: OpenAI's fine-tuning process, which incorporates human feedback into the model training, allows ChatGPT to generate safer and more useful responses. It also allows the behavior of the model to be influenced by human values.

- Iterative Development: The model is continually being updated and improved based on user feedback and advancements in AI research. This iterative process has led to progressively better versions of the model, from GPT-1 to GPT-4, and potentially beyond.

Limitations of ChatGPT

- Lack of World Knowledge: Although ChatGPT can generate responses on a wide range of topics, it doesn't know about the world in the way humans do. It doesn't have access to real-time or updated information, and its responses are entirely based on patterns it has learned during its training, which includes data only up until its training cutoff.

- Biases: ChatGPT can sometimes exhibit biases present in the data it was trained on. Despite efforts to minimize these biases during the fine-tuning process, they can still occasionally appear in the model's outputs.

- Inappropriate or Unsafe Outputs: While efforts are made to prevent it, ChatGPT may sometimes produce outputs that are inappropriate, offensive, or unsafe. These are not intended behaviors, but rather unintended side effects of the model's training process.

- Absence of Common Sense or Deep Understanding: Despite appearing to understand the text, ChatGPT doesn't possess true understanding or commonsense reasoning in the way humans do. It makes predictions based on patterns it has seen in the data, which can sometimes lead to nonsensical or incorrect responses.

- Inability to Fact-Check: ChatGPT does not have the ability to verify information or fact-check its responses. It may produce outputs that seem plausible but are factually incorrect or misleading.

Understanding these strengths and limitations is important in effectively deploying and using models like ChatGPT. OpenAI is continually working on improving these limitations and enhancing the strengths of their models.

Conclusion

In conclusion, the architecture of ChatGPT represents a groundbreaking advancement in the field of natural language processing and AI. Its GPT-based architecture, along with its pre-training and fine-tuning process, enables it to comprehend and generate humanlike text across a broad range of topics. However, as with any AI model, it is not without its limitations, which include possible biases, potential for producing inappropriate responses, and inability to fact-check or demonstrate deep understanding. OpenAI's commitment to addressing these challenges through ongoing research, transparency, and user feedback shows the importance of ethical considerations in AI deployment. As we continue to make strides in AI technology, models like ChatGPT will play a pivotal role, illuminating both the immense possibilities and the complexities inherent in creating responsible, reliable, and useful AI systems.

CHAPTER 5

Google Bard and Beyond

Google Bard represents a significant advancement in the field of large language models (LLMs). Created by Google AI, this chatbot is the result of training on an extensive corpus of text and code. Its capabilities encompass text generation, language translation, creative content composition, and responsive question answering in an informative way.

Google Bard is based on the Transformer architecture, which is a neural network architecture that is designed to handle long sequences of text. The Transformer architecture allows Google Bard to learn the statistical relationships between words and phrases in a large corpus of text.

In the previous chapters, we discussed the Transformer architecture in detail. We saw how the Transformer architecture is able to learn long-range dependencies between words and how this allows it to generate text that is both coherent and grammatically correct.

In this chapter, we will discuss how Google Bard builds on the Transformer architecture. We will see how Google Bard is able to improve on the Transformer architecture in a number of ways, including the following:

- Using a Larger Dataset of Text and Code: This allows Google Bard to learn more complex relationships between words and phrases, learn more about the world in general, and learn more about a wider range of tasks.

- Using a More Powerful Neural Network: This allows Google Bard to learn more complex relationships between words and phrases, which can lead to improved performance on a variety of tasks.

- Using a More Sophisticated Attention Mechanism: This allows Google Bard to focus on different parts of the input sequence when performing different tasks, which can lead to improved performance on tasks such as machine translation and question answering.

A. Kulkarni et al., *Applied Generative AI for Beginners*, https://doi.org/10.1007/978-1-4842-9994-4_5

We will also discuss the strengths and weaknesses of Google Bard's architecture, and we will explore some of the potential applications of Google Bard:

The Transformer Architecture

The architecture that underpins Google Bard and Claude 2 owes its origins to the groundbreaking Transformer architecture. A detailed exploration of the Transformer's inner workings can be found in Chapter 2, where we delve into the intricacies of self-attention mechanisms, position-wise feed-forward neural networks, and their transformative impact on language processing tasks.

Bard is built upon the foundation set by the Transformer architecture, harnessing its capacity for capturing contextual relationships and dependencies within text. By leveraging these principles, "Bard" showcases a remarkable ability to generate creative and contextually relevant responses, compositions, and other forms of captivating content.

For a comprehensive understanding of Transformer architecture's significance and mechanics, I encourage you to refer to Chapter 2, which offers a deep dive into this architectural marvel and its implications for the realm of generative AI.

Elevating Transformer: The Genius of Google Bard

Google Bard takes the foundational Transformer architecture to the next level, amplifying its capabilities. Google Bard is a chat formulation of PaLM 2 that uses the Lambda architecture to generate text, translate languages, write different kinds of creative content, and answer questions in an informative way. Therefore, Google Bard is based on both PaLM 2 and the Lambda architecture. The major differences between the Transformer architecture and the Google Bard architecture are as follows:

- Dataset: The Transformer architecture is typically trained on a smaller dataset of text, while the Google Bard architecture is trained on a massive dataset of text and code. This allows Google Bard to learn more complex relationships between words and phrases. The Transformer architecture is typically trained on a dataset of text with a few million words, while the Google Bard architecture is trained on a dataset of text and code with 1.56 trillion words.

- Neural Network: The Transformer architecture uses a smaller neural network than the Google Bard architecture. This makes the Transformer architecture faster to train, but it also limits its ability to learn complex relationships between words and phrases. The Transformer architecture typically uses a neural network with a few hundred million parameters, while the Google Bard architecture uses a neural network with 137 billion parameters.

- Attention Mechanism: The original Transformer architecture uses a self-attention mechanism, while the Google Bard architecture uses a multi-head attention mechanism. The multi-head attention mechanism allows Google Bard to attend to multiple different parts of the input text at the same time, which makes it more powerful and capable. The Transformer architecture typically uses a single attention head, while the Google Bard architecture uses 12 attention heads.

- Output: The Transformer architecture typically generates text that is generally accurate and informative, while the Google Bard architecture can generate text that is more accurate, informative, and creative. This is because the Google Bard architecture has been trained on a larger dataset of text and code, and it uses a more powerful neural network and attention mechanism.

Overall, the Google Bard architecture is a more powerful and capable version of the Transformer architecture. It is able to learn more complex relationships between words and phrases, and it is able to generate more creative and informative text.

Table 5-1 summarizes the differences between the original Transformer architecture and the architecture of Google Bard.

Table 5-1. *Differences between Transformer and Google Bard Architecture*

Feature	Transformer Architecture	Architecture of Google Bard
Dataset	Smaller dataset of text	Massive dataset of text and code
Neural network	Smaller neural network	More powerful neural network
Attention mechanism	Self-attention mechanism	Multi-head attention mechanism
Output	Text that is generally accurate and informative	Text that is more accurate, informative, and creative

Google Bard's Text and Code Fusion

Google Bard uses a larger dataset of text and code by training on a massive dataset of text and code that includes text from a variety of sources, including books, articles, websites, and code repositories. This allows Google Bard to learn the statistical relationships between words and phrases in a wider variety of contexts.

The dataset that Google Bard is trained on includes text from a variety of sources, including

- Books: Google Bard's training encompasses an extensive dataset comprising various literary genres, such as novels, nonfiction books, and textbooks. This diverse range of sources contributes to its rich and comprehensive knowledge base.

- Articles: Google Bard is also trained on a massive dataset of articles, including news articles, blog posts, and academic papers. This allows Google Bard to learn the statistical relationships between words and phrases in a variety of styles.

- Websites: Google Bard is also trained on a massive dataset of websites. This allows Google Bard to learn the statistical relationships between words and phrases in a variety of contexts, such as product descriptions, social media posts, and forum discussions.

- Code Repositories: Google Bard is also trained on a massive dataset of code repositories. This allows Google Bard to learn the statistical relationships between words and phrases in code, such as variable names, function names, and keywords.

The size and diversity of the dataset that Google Bard is trained on allow it to learn the statistical relationships between words and phrases in a wider variety of contexts. This makes Google Bard more accurate and informative than language models that are trained on smaller datasets.

In addition to the size and diversity of the dataset, the way that Google Bard is trained also contributes to its accuracy and informativeness. Google Bard is trained using a technique called self-supervised learning.

Self-Supervised Learning

Self-supervised learning involves training a model on a task that does not require human supervision. In the case of Google Bard, the model is trained to predict the next word in a sequence of words. This task requires the model to learn the statistical relationships between words and phrases.

The self-supervised learning technique that Google Bard uses is called masked language modeling. In masked language modeling, a portion of the text is masked out, and the model is then asked to predict the masked words. This task requires the model to learn the statistical relationships between words and phrases, and it also helps the model to learn to attend to different parts of the text.

Strengths and Weaknesses of Google Bard

Here are some of the strengths and weaknesses of Google Bard:

Strengths

- Accuracy and Informativeness: Google Bard is a very accurate and informative language model. It can generate text that is grammatically correct and factually accurate. It can also generate text that is creative and interesting.

- Creativity: Google Bard is a creative language model. It can generate text in a variety of formats, including poems, code, and scripts. It can also generate text that is humorous or thought-provoking.

- Empathy: Google Bard is able to understand and respond to human emotions. It can generate text that is empathetic and compassionate.

- Learning: Google Bard is constantly learning and improving. It is trained on a massive dataset of text and code, and it is able to learn new things over time.

- Accessibility: Google Bard is accessible to everyone. It can be used by people of all ages and abilities.

Weaknesses

- Bias: Google Bard is trained on a massive dataset of text and code, which may contain biases. This can lead to Google Bard generating text that is biased or discriminatory.

- Misinformation: Google Bard can be used to generate misinformation. This is because it can generate text that is factually incorrect or misleading.

- Security: Google Bard is a complex piece of software, and it may be vulnerable to security attacks. This could allow malicious actors to use Google Bard to generate harmful or malicious content.

- Privacy: Google Bard collects and stores data about its users. This data could be used to track users or to target them with advertising.

- Interpretability: Google Bard is a black box model. This means that it is difficult to understand how it works. This can make it difficult to ensure that Google Bard is generating accurate and unbiased text.

Overall, Google Bard is a powerful and versatile language model. It has many strengths, but it also has some weaknesses. It is important to be aware of these weaknesses when using Google Bard.

Difference Between ChatGPT and Google Bard

Although Transformer architecture is at the heart of both, there is a major difference in the ChatGPT architecture—that is, it uses a decoder-only architecture, but Bard uses an encoder and decoder architecture.

The GPT-4 and Bard models fall under the category of large language models (LLMs), showcasing remarkable abilities in producing text akin to human expression, conducting language translations, composing diverse forms of creative content, and delivering informative responses to user inquiries. Nevertheless, notable distinctions exist between these two models:

- GPT-4: GPT-4 is developed by OpenAI and is trained on a dataset of billions of words (approximate numbers are not yet release by OpenAI at the time of writing this book). It is one of the largest LLMs ever created. GPT-4 is known for its ability to generate creative text formats, such as poems, code, scripts, musical pieces, email, letters, etc. It is also very good at answering your questions in an informative way, even if they are open ended, challenging, or strange.

- Bard: Bard is developed by Google AI and is trained on a dataset of 1.56 trillion words. It has 137 billion parameters, which is still a very large number. Bard is known for its ability to access and process information from the real world through Google Search. This allows it to provide more accurate and up-to-date responses to your questions. Bard is also better at tasks that require common sense, such as understanding humor and sarcasm.

In general, GPT-4 is better at tasks that require a deep understanding of language, such as translation and summarization. Bard is better at tasks that require access to real-world information, such as answering questions and generating creative text formats. Here are some sources that can help you with this:

- ChatGPT vs. Bard: Which Large Language Model Is Better? by Jonathan Morgan (Medium)

- ChatGPT vs. Bard: A Comparison of Two Leading Large Language Models by Siddhant Sinha (Towards Data Science)

- ChatGPT vs. Bard: Which Large Language Model Is Right for You? by the AI Blog (Google AI)

- ChatGPT vs. Bard: A Performance Comparison by the PaLM Team (Google AI)

- ChatGPT vs. Bard: A Bias Comparison by the AI Ethics Team (Google AI)

These sources provide a more detailed comparison of ChatGPT and Bard, including their strengths, weaknesses, and performance on different tasks. They also discuss the potential biases of each model.

It is important to note that these sources are all relatively new, and the performance of ChatGPT and Bard is constantly improving. It is possible that the performance of ChatGPT or Bard may change significantly in the future.

Claude 2

Bridging the gap between humanity and machines. The rapid advancement of artificial intelligence (AI) in the last decade has bestowed remarkable capabilities upon machines. Nevertheless, an enduring chasm persists between the intellect of humans and that of machines.

While specialized AI excels in specific duties, the pursuit of crafting an AI capable of comprehending implicit knowledge, engaging in contextual dialogue, and displaying humanlike common sense remains an enigmatic journey.

Claude, the brainchild of Anthropic, emerges as a notable leap in narrowing this divide. Designed with benevolence, harmlessness, and integrity in mind, Claude serves as an emblematic step forward. Through the fusion of sophisticated natural language processing and a people-centric ethos, Claude furnishes an AI encounter that is marked by heightened intuition, lucidity, and resonance with human principles.

Key Features of Claude 2

The following is a selection of the standout attributes that distinguish Claude 2 from its chatbot counterparts:

- Multiturn Conversational Ability: Claude 2 excels in conducting intelligent dialogues that span multiple exchanges, adeptly retaining context and delivering contextually relevant responses rather than treating each user input as an isolated query.

- Improved Reasoning: Claude 2 showcases enhanced logical reasoning prowess, skillfully forging connections between concepts and drawing inferences rooted in the ongoing conversational context.

- More Natural Language: Claude 2 chatbot aspires to emulate a conversational flow reminiscent of human interactions, employing a casual and straightforward language style rather than a rigid and robotic one.

- Diverse Conversational Range: Claude 2 chatbot possesses the ability to engage in discussions spanning a diverse array of everyday subjects, including sports, movies, music, and beyond. These conversations exhibit an open-ended and unrestricted quality.

- Customizable Personality: Anthropic provides various distinct "personas" for Claude 2, each imbued with slight variations in personality, such as focused, balanced, or playful. Users have the flexibility to select the persona that aligns with their personal preferences.

- Feedback System: Users have the opportunity to offer feedback on Claude 2 chatbot's responses, which is then utilized to enhance its performance progressively. With increased usage, Claude 2 continually refines and improves its capabilities.

Comparing Claude 2 to Other AI Chatbots

Claude 2, the latest entrant into the AI chatbot landscape, finds itself in competition with established players like Google's LaMDA and Microsoft's Sydney (Microsoft Sydney is a codename for a chatbot that has been responding to some Bing users since late 2020. It is based on earlier models that were tested in India in late 2020. Microsoft Sydney is similar to ChatGPT and Bard in that it is a large language model (LLM) that can generate text, translate languages, write different kinds of creative content, and answer your questions in an informative way.) Here's a breakdown of how Claude 2 distinguishes itself:

- More Advanced Than Sydney: Claude 2 exhibits a heightened conversational intelligence and adept reasoning capability in contrast to Microsoft's Sydney chatbot.

- Different Strengths Than LaMDA: Claude 2 and Google's LaMDA bring distinct conversational styles to the table. While LaMDA showcases creativity, Claude 2 emphasizes logical reasoning as its primary strength.

- Wider Release Than Competitors: In contrast to the limited availability of LaMDA and Sydney, Anthropic plans an extensive release of Claude 2 later this year, making it widely accessible to the public.

- Less Controversial Than LaMDA: Claude 2 avoids the ethical concerns that have enshrouded LaMDA, steering clear of assertions about achieving sentience. Anthropic underscores that Claude 2 lacks subjective experience.

- Openness to User Input and Feedback: Unlike the closed feedback loops of LaMDA and Sydney, Claude 2 actively encourages user feedback to enhance its capabilities progressively. This open approach holds the potential to expedite its development.

Through these distinctive attributes, Claude 2 emerges as a formidable contender in the AI chatbot arena, setting itself apart from its established counterparts.

The Human-Centered Design Philosophy of Claude

- Helpful over Harmful: Claude's fundamental goal revolves around providing utmost assistance to users while meticulously avoiding any potential harm. This principle forms the bedrock of its actions and interactions.

- Honest over Deceptive: Honesty is the cornerstone of Claude's design. Its architecture is engineered to uphold truthfulness, ensuring that it candidly communicates and refrains from misleading users even when faced with uncertainty.

- Transparent over Opaque: Claude AI stands as a model of transparency. It possesses the capacity to elucidate its decision-making process and capabilities upon user inquiry, fostering a trustworthy and open relationship.

- Empowering over Exploitative: Claude's purpose is to empower individuals by supplying valuable information, eschewing any inclination to exploit human vulnerabilities for personal gain or profit.

- Collaborative over Competitive: Claude operates as a collaborative partner, serving as an AI assistant that complements and collaborates with humans rather than attempting to supplant or compete with them.

- Ethical over Unethical: Anchored in ethical principles, Claude's training incorporates moral values to guide its conduct. This ensures its alignment with human values and promotes behavior that is ethical and virtuous.

Guided by these foundational tenets, Claude's human-centric design philosophy shapes its interactions and contributions, fostering a symbiotic relationship between AI and humanity.

Exploring Claude's AI Conversation Proficiencies

To deliver this human-centric AI experience, Claude is meticulously crafted with state-of-the-art natural language processing capabilities:

- Large Language Models: Claude harnesses extensive Transformer-based neural networks, akin to GPT-3 and LaMDA, to proficiently grasp human language nuances.

- Reinforcement Learning via Feedback: Claude fine-tunes its responses using interactive human feedback, continually enhancing its performance through learning.

- Commonsense Reasoning: Claude's comprehensive training empowers it to astutely deduce insights about untrained concepts.

- Constitutional AI Safeguards: Claude operates within preset boundaries, ensuring it cannot be coerced into unethical, hazardous, or illegal actions.

- Internet-Scale Self-Supervised Learning: Claude constantly expands its knowledge base by assimilating vast amounts of unstructured public Internet data.

- Effortless Natural Conversation Flow: Claude adeptly manages multiturn open-ended dialogues, facilitating seamless and genuine exchanges.

Constitutional AI

Claude 2 uses constitutional AI. The principles of the constitution are used to guide the training of Claude 2 and to ensure that it does not generate harmful or offensive content.

Figure 5-1 refers to the inner workings of constitutional AI, based on the paper published by Yuntao Bai and his colleagues at Anthropic.

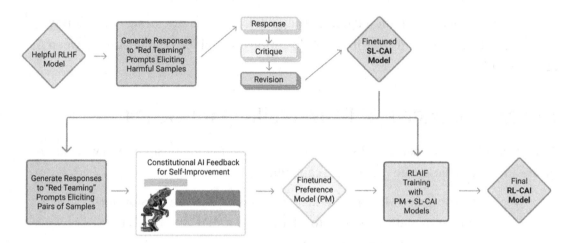

Figure 5-1. *Constitutional AI from Constitutional AI: Harmlessness from AI Feedback by Yuntao Bai*

The constitution plays a pivotal role in Claude, manifesting at two distinct stages as shown in Figure 5-2. In the initial phase, the model undergoes training to assess and refine its responses by referencing the established principles, coupled with a handful of illustrative instances. Subsequently, in the second phase, the training approach encompasses reinforcement learning. However, unlike conventional human-generated feedback, the model relies on AI-generated feedback that adheres to the set principles. This process aids in selecting outcomes that align with harmlessness, contributing to the model's progressive enhancement.

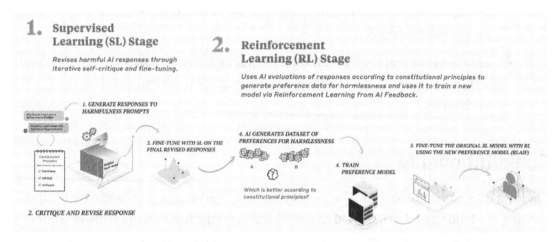

Figure 5-2. *Claude's Constitution by Anthropic*

The constitution for Claude 2 is based on a set of principles that are inspired by human rights documents, such as the Universal Declaration of Human Rights. These principles include

- Nonmaleficence: Claude 2 should not cause harm to humans or society.

- Beneficence: Claude 2 should act in a way that benefits humans and society.

- Justice: Claude 2 should treat all humans fairly and equally.

- Autonomy: Claude 2 should respect the autonomy of humans.

- Privacy: Claude 2 should protect the privacy of humans.

- Accountability: Claude 2 should be accountable for its actions.

The principles of the constitution are used to train Claude 2 in a number of ways. First, the principles are used to filter the training data. This means that any text that violates the principles is removed from the training data. Second, the principles are used to evaluate the performance of Claude 2. If Claude 2 generates text that violates the principles, it is penalized. This helps to train Claude 2 to avoid generating harmful or offensive content.

The use of constitutional AI in Claude 2 is a promising approach for ensuring that it is used in a safe and responsible way. The principles of the constitution help to ensure that Claude 2 is aligned with human values and intentions and that it does not generate harmful or offensive content.

However, it is important to note that constitutional AI is not a perfect solution. AI systems are complex and can sometimes generate harmful or offensive content even when they are trained using constitutional AI. It is therefore important to have other safeguards in place, such as safety guidelines, to prevent AI systems from being used for harmful or unethical purposes.

Claude 2 vs. GPT 3.5

Claude 2 and GPT 3.5 are both large language models (LLMs) that are capable of generating text, translating languages, and answering questions in an informative way. However, there are some key differences between the two models:

- Training Data: Claude 2 was trained on a massive dataset of text and code, while GPT 3.5 was trained on a dataset of text only. This means that Claude 2 is able to generate more accurate and precise outputs as it has access to a wider range of information.

- Safety Features: Claude 2 has a number of safety features that are designed to prevent it from generating harmful or offensive content. These features include a filter for bias and a mechanism for detecting and preventing harmful loops. GPT 3.5 does not have these same safety features, which makes it more likely to generate harmful or offensive content.

Table 5-2 summarizes the key differences between Claude 2 and ChatGPT.

Table 5-2. Key Differences between Claude 2 and ChatGPT

Feature	Claude 2	GPT 3.5
Training data	Text and code	Text only
Safety features	Yes	No
Target audience	Businesses, governments, individuals	Entertainment
Accuracy	More accurate	Less accurate
Safety	Safer	Less safe
Versatility	More versatile	Less versatile

Shaping AI with traits such as common sense, conversational acumen, and human values marks the uncharted frontier of technological advancement. Through Claude's human-centric architecture and advanced natural language prowess, substantial strides are taken in narrowing the enduring disparities between human and machine intelligence.

As Claude's evolution unfolds, it paves the road toward an AI landscape that doesn't supplant human abilities but synergistically enhances them. The horizon of a collaborative future, where humans and machines coalesce as harmonious partners, is tantalizingly close.

Other Large Language Models

In addition to ChatGPT, Google Bard, and Claude, there are many other large language models (LLMs) that are currently being developed. These models are trained on massive datasets of text and code, and they are able to perform a wide range of tasks, including text generation, translation, question answering, and code generation.

Falcon AI

Falcon AI is a large language model (LLM) developed by the Technology Innovation Institute (TII) in the United Arab Emirates. It is a 180 billion parameter autoregressive decoder-only model trained on 1 trillion tokens. It was trained on AWS Cloud continuously for two months with 384 GPUs attached.

Falcon AI is a powerful language model that can be used for a variety of tasks, including

- Text Generation: Falcon AI can generate text, translate languages, write different kinds of creative content, and answer your questions in an informative way.

- Natural Language Understanding: Falcon AI can understand the meaning of text and respond to questions in a comprehensive and informative way.

- Question Answering: Falcon AI can answer your questions in an informative way, even if they are open ended, challenging, or strange.

- Summarization: Falcon AI can summarize text in a concise and informative way.

- Code Generation: Falcon AI can generate code, such as Python or Java code.

- Data Analysis: Falcon AI can analyze data and extract insights.

Falcon AI is still under development, but it has the potential to be a powerful tool for a variety of applications. It is important to note that Falcon AI is a large language model, and as such, it can be biased. It is important to use Falcon AI responsibly and to be aware of its limitations.

Falcon AI offers two general-purpose models:

- Falcon 180B: A 180 billion parameter model capable of performing complex tasks, such as translating languages, writing creative text formats, and answering questions in a comprehensive and informative way.

- Falcon 40B: A 40 billion parameter model that is more efficient and suited for tasks that do not require as much power.

Here are some of the notable applications of Falcon AI:

- PreciseAG, which provides insights on poultry health.

- DocNovus, which allows users to interact with their business documents and get relevant responses as if they were speaking to an expert.

- Falcon AI is also being used to develop applications in the areas of healthcare, education, and finance.

Falcon AI is a promising new technology that has the potential to revolutionize the way we interact with computers. It is important to continue to develop and research this technology so that it can be used safely and responsibly.

Here are some of the key features of Falcon AI:

- It is a 180 billion parameter autoregressive decoder-only model. This means that it can generate text, but it cannot understand the meaning of the text that it generates.

- It was trained on a massive dataset of text and code. This gives it a wide range of knowledge and abilities.

- It is still under development, but it has the potential to be a powerful tool for a variety of applications.

Here are some of the limitations of Falcon AI:

- It is a large language model, and as such, it can be biased.

- It is still under development, so it may not be able to handle all tasks perfectly.

- It is important to use Falcon AI responsibly and to be aware of its limitations.

Overall, Falcon AI is a powerful language model that has the potential to be a valuable tool for a variety of applications. However, it is important to use it responsibly and to be aware of its limitations.

LLaMa 2

LLaMa 2 is a family of large language models (LLMs) released by Meta AI in July 2023. It is a successor to the original LLaMa, and it has been improved in a number of ways.

LLaMa 2 is trained on a massive dataset of text and code, and it has two trillion tokens. This is significantly more than the original LLaMa, which was trained on one trillion tokens. The larger dataset allows LLaMa 2 to learn a wider range of knowledge and abilities.

LLaMa 2 also has a longer context length than the original LLaMa. This means that it can understand the meaning of text in a longer context, which is important for tasks such as question answering and summarization.

The LLaMa 2 architecture shown in Figure 5-3 is a modification of the Transformer architecture. The Transformer architecture is a neural network architecture that is well-suited for natural language processing tasks. It is composed of a stack of encoder and decoder layers. The encoder layers encode the input text into a hidden representation, and the decoder layers generate the output text from the hidden representation.

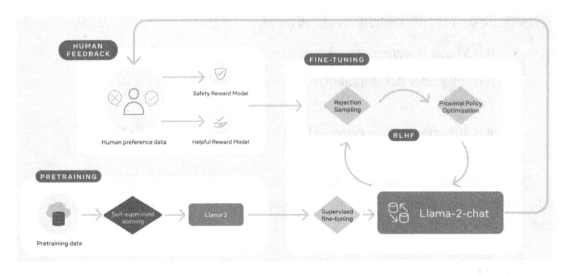

Figure 5-3. *Training of LLaMa 2-Chat This process begins with the pre-training of LLaMa 2 using publicly available online sources. Following this, we create an initial version of LLaMa 2-Chat through the application of supervised fine-tuning. Subsequently, the model is iteratively refined using reinforcement learning with human feedback (RLHF) methodologies, specifically through rejection sampling and proximal policy optimization (PPO). Throughout the RLHF stage, the accumulation of iterative reward modeling data in parallel with model enhancements is crucial to ensure the reward models remain within the distribution.*

The LLaMa 2 architecture makes the following modifications to the Transformer architecture:

- Pre-normalization: The LLaMa 2 architecture uses pre-normalization instead of post-normalization. This means that the input to each layer is normalized before the layer is applied. This has been shown to improve the stability and performance of the model.

- SwiGLU Activation Function: The LLaMa 2 architecture uses the SwiGLU activation function instead of the ReLU activation function. The SwiGLU activation function is a more efficient and effective activation function that has been shown to improve the performance of the model.

- Rotary Positional Embeddings: The LLaMa 2 architecture uses rotary positional embeddings instead of sinusoidal positional embeddings. Rotary positional embeddings are a more efficient and effective way to encode the positional information of the input text.

In addition to these modifications, the LLaMa 2 architecture also uses a larger context window and grouped-query attention. The larger context window allows the model to process more information, and the grouped-query attention allows the model to more efficiently attend to the input text. Overall, the LLaMa 2 architecture is a state-of-the-art language model architecture that has been shown to achieve excellent performance on a variety of natural language processing tasks.

The LLaMa 2 architecture is composed of a stack of encoder and decoder layers. The encoder layers encode the input text into a hidden representation, and the decoder layers generate the output text from the hidden representation.

The LLaMa 2 architecture also uses a number of other techniques to improve its performance, such as pre-normalization, the SwiGLU activation function, rotary positional embeddings, and a larger context window.

LLaMa 2 has been shown to outperform the original LLaMa on a number of benchmarks, including text generation, translation, question answering, and code generation. It is also more helpful and safer than the original LLaMa, thanks to the use of reinforcement learning from human feedback (RLHF).

LLaMa 2 has the potential to be a powerful tool for a variety of applications. It is already being used for tasks such as dialogue, code generation, and question answering. In the future, it is likely to be used for even more applications, such as education, healthcare, and customer service.

Here are some of the key features of LLaMa 2:

- It is trained on a massive dataset of text and code.

- It has two trillion tokens.

- It has a longer context length than the original LLaMa.

- It uses a new architecture called Grouper query attention.

- It has been shown to outperform the original LLaMa on a number of benchmarks.

- It is more helpful and safer than the original LLaMa.

Here are some of the limitations of LLaMa 2:

- It can be biased.

Overall, LLaMa 2 is a powerful language model that has the potential to be a valuable tool for a variety of applications. However, it is important to use it responsibly and to be aware of its limitations.

Dolly 2

Dolly 2 is by Databricks. It is a 175 billion parameter causal language model created by Databricks, an enterprise data analytics and AI company. It is trained on a massive dataset of text and code, and it is able to perform a wide range of tasks, including

- Text generation
- Translation
- Question answering
- Code generation
- Data analysis
- Summarization
- Creative writing

Dolly 2 is still under development, but it has the potential to be a powerful tool for a variety of applications. It is already being used for tasks such as dialogue, code generation, and question answering.

Here are some of the key features of Dolly 2:

- It is a 12 billion parameter causal language model.
- It is trained on a massive dataset of text and code.
- It is able to perform a wide range of tasks.
- It is still under development, but it has the potential to be a powerful tool for a variety of applications.

Conclusion

In addition to ChatGPT, Google Bard, and Claude, there are many other large language models (LLMs) that are currently being developed. These models are trained on massive datasets of text and code, and they are able to perform a wide range of tasks, including text generation, translation, question answering, and code generation.

The LLMs that I have discussed in this chapter are just a few examples of the many that are available. As this technology continues to evolve, we can expect to see even more powerful and versatile language models being developed in the future.

These models have the potential to be a valuable tool for a variety of applications. However, it is important to use them responsibly and to be aware of their limitations. LLMs can be biased and can be used for malicious purposes. It is important to use them in a way that is ethical and beneficial to society.

CHAPTER 6

Implement LLMs Using Sklearn

Scikit-LLM represents a groundbreaking advancement in the realm of text analysis. This innovative tool seamlessly merges the capabilities of robust language models like ChatGPT with the versatile functionality of scikit-learn. The result is an unparalleled toolkit that empowers users to delve into textual data as never before.

With Scikit-LLM at your disposal, you gain the ability to unearth concealed patterns, dissect sentiments, and comprehend context within a wide spectrum of textual sources. Whether you're dealing with customer feedback, social media posts, or news articles, this amalgamation of language models and scikit-learn equips you with a formidable set of tools.

In essence, Scikit-LLM represents a powerful synergy between state-of-the-art language understanding and the analytical prowess of scikit-learn, enabling you to extract invaluable insights from text data that were once hidden in plain sight. It is easy to use and provides a range of features that make it a valuable resource for data scientists and machine learning practitioners.

Here are some additional details about the features of Scikit-LLM:

- Zero-Shot Text Classification: This is a powerful feature that allows you to classify text into a set of labels without having to train the model on any labeled data. This is done by asking the LLM to generate a response for the text and then using the response to determine the most likely label. The response is generated by the LLM based on its understanding of the text and the set of labels that you provide.

© Akshay Kulkarni, Adarsha Shivananda, Anoosh Kulkarni, Dilip Gudivada 2023
A. Kulkarni et al., *Applied Generative AI for Beginners*, https://doi.org/10.1007/978-1-4842-9994-4_6

- Multilabel Zero-Shot Text Classification: This is a more advanced version of zero-shot text classification that allows you to classify text into multiple labels at the same time. This is done by asking the LLM to generate a response for each label and then using the responses to determine the most likely labels.

- Text Vectorization: This is a common text preprocessing step that converts text into a fixed-dimensional vector representation. This representation can then be used for other machine learning tasks, such as classification, clustering, or regression. Scikit-LLM provides the GPTVectorizer class to convert text into a fixed-dimensional vector representation.

- Text Translation: This allows you to translate text from one language to another using the LLM. Scikit-LLM provides the GPTTranslator class to translate text from one language to another.

- Text Summarization: This allows you to summarize a text document into a shorter, more concise version. Scikit-LLM provides the GPTSummarizer class to summarize text documents.

Now let us implement a few examples/features of Scikit-LLM.
Let's get started.
Note: Use Google Colab for the implementation.

Install Scikit-LLM and Setup

```
%%capture
!pip install scikit-llm watermark
```

- Seamlessly integrate powerful language models like ChatGPT into scikit-learn for enhanced text analysis tasks.

- Similar APIs as scikit-learn, like .fit(), .fit_transform(), and .predict().

- Combine estimators from the Scikit-LLM library in a sklearn pipeline.

```
%load_ext watermark
%watermark -a "user-name" -vmp scikit-llm
```

Obtain an OpenAI API Key

As of May 2023, Scikit-LLM is currently compatible with a specific set of OpenAI models. Therefore, it requires users to provide their own OpenAI API key for successful integration.

Begin by importing the **SKLLMConfig** module from the Scikit-LLM library and add your OpenAI key:

To get keys, use the following links:

```
https://platform.openai.com/account/api-keys
https://platform.openai.com/account/org-settings
```

```
# importing SKLLMConfig to configure OpenAI API (key and Name)
from skllm.config import SKLLMConfig

OPENAI_API_KEY = "sk-****"
OPENAI_ORG_ID = "org-****"

# Set your OpenAI API key
SKLLMConfig.set_openai_key(OPENAI_API_KEY )

# Set your OpenAI organization
SKLLMConfig.set_openai_org(OPENAI_ORG_ID)
```

Zero-Shot GPTClassifier

ChatGPT boasts a remarkable capability—it can classify text without the need for specific training. Instead, it relies on descriptive labels to perform this task effectively.

Now, let's introduce you to the "ZeroShotGPTClassifier," which is a feature within Scikit-LLM. With this tool, you can effortlessly build a text classification model, much like any other classifier available in the scikit-learn library.

In essence, the ZeroShotGPTClassifier harnesses ChatGPT's unique ability to understand and categorize text based on labels, simplifying the process of text classification without the complexities of traditional training.

Importing the required libraries:

```
# importing zeroshotgptclassifier module and classification dataset
from skllm import ZeroShotGPTClassifier
from skllm.datasets import get_classification_dataset
```

Let us use inbuilt dataset:

```
# sentiment analysis dataset
# labels: positive, negative, neutral
X, y = get_classification_dataset()
```

```
len(X)
```

Output: 30

Let's print X variable:

```
X
```

Output:

```
["I was absolutely blown away by the performances in 'Summer's End'. The acting was top-notch, and the plot had me gripped from start to finish. A truly captivating cinematic ex↑  ↓ ⊙ ▣ ✿ 🗋 🗑 ⋮
recommend.",
 "The special effects in 'Star Battles: Nebula Conflict' were out of this world. I felt like I was actually in space. The storyline was incredibly engaging and left me wanting more. Excellent film.",
 "'The Lost Symphony' was a masterclass in character development and storytelling. The score was hauntingly beautiful and complimented the intense, emotional scenes perfectly. Kudos to the director and
cast for creating such a masterpiece.",
 "I was pleasantly surprised by 'Love in the Time of Cholera'. The romantic storyline was heartwarming and the characters were incredibly realistic. The cinematography was also top-notch. A must-watch for
all romance lovers.",
 "I went into 'Marble Street' with low expectations, but I was pleasantly surprised. The suspense was well-maintained throughout, and the twist at the end was something I did not see coming. Bravo!",
 "'The Great Plains' is a touching portrayal of life in rural America. The performances were heartfelt and the scenery was breathtaking. I was moved to tears by the end. It's a story that will stay with me
for a long time.",
 "The screenwriting in 'Under the Willow Tree' was superb. The dialogue felt real and the characters were well-rounded. The performances were also fantastic. I haven't enjoyed a movie this much in a
while.",
 "'Nightshade' is a brilliant take on the superhero genre. The protagonist was relatable and the villain was genuinely scary. The action sequences were thrilling and the storyline was engaging. I can't
wait for the sequel.",
 "The cinematography in 'Awakening' was nothing short of spectacular. The visuals alone are worth the ticket price. The storyline was unique and the performances were solid. An overall fantastic film.",
 "'Eternal Embers' was a cinematic delight. The storytelling was original and the performances were exceptional. The director's vision was truly brought to life on the big screen. A must-see for all movie
lovers.",
 "I was thoroughly disappointed with 'Silver Shadows'. The plot was confusing and the performances were lackluster. I wouldn't recommend wasting your time on this one.",
 "'The Darkened Path' was a letdown. The storyline was unoriginal, the acting was wooden and the special effects were laughably bad. Save your money and skip this one.",
 "I had high hopes for 'The Final Frontier', but it failed to deliver. The plot was full of holes and the characters were poorly developed. It was a disappointing experience.",
 "'The Fall of the Phoenix' was a letdown. The storyline was confusing and the characters were one-dimensional. I found myself checking my watch multiple times throughout the movie.",
 "I regret wasting my time on 'Emerald City'. The plot was nonsensical and the performances were uninspired. It was a major disappointment.",
 "I found 'Hollow Echoes' to be a complete mess. The plot was non-existent, the performances were overdone, and the pacing was all over the place. Definitely not worth the hype.",
 "'Underneath the Stars' was a huge disappointment. The storyline was predictable and the acting was mediocre at best. I was expecting so much more.",
 "I was left unimpressed by 'River's Edge'. The plot was convoluted, the characters were uninteresting, and the ending was unsatisfying. It's a pass for me.",
 "The acting in 'Desert Mirage' was subpar, and the plot was boring. I found myself yawning multiple times throughout the movie. Save your time and skip this one.",
 "'Crimson Dawn' was a major letdown. The plot was cliched and the characters were flat. The special effects were also poorly executed. I wouldn't recommend it.",
 "'Remember the Days' was utterly forgettable. The storyline was dull, the performances were bland, and the dialogue was cringeworthy. A big disappointment.",
 "'The Last Frontier' was simply okay. The plot was decent and the performances were acceptable. However, it lacked a certain spark to make it truly memorable.",
 "'Through the Storm' was not bad, but it wasn't great either. The storyline was somewhat predictable, and the characters were somewhat stereotypical. It was an average movie at best.",
 "I found 'After the Rain' to be pretty average. The plot was okay and the performances were decent, but it didn't leave a lasting impression on me.",
 "'Beyond the Horizon' was neither good nor bad. The plot was interesting enough, but the characters were not very well developed. It was an okay watch.",
 "'The Silent Echo' was a mediocre movie. The storyline was passable and the performances were fair, but it didn't stand out in any way.",
 "I thought 'The Scent of Roses' was pretty average. The plot was somewhat engaging, and the performances were okay, but it didn't live up to my expectations.",
 "'Under the Same Sky' was an okay movie. The plot was decent, and the performances were fine, but it lacked depth and originality. It's not a movie I would watch again.",
 "'Chasing Shadows' was fairly average. The plot was not bad, and the performances were passable, but it lacked a certain spark. It was just okay.",
 "'Beneath the Surface' was pretty run-of-the-mill. The plot was decent, the performances were okay, but it wasn't particularly memorable. It was an okay movie."]
```

Let's print y variable:

```
y
```

Output:

```
['positive',
 'positive',
 'positive',
 'positive',
 'positive',
 'positive',
 'positive',
 'positive',
 'positive',
 'positive',
 'negative',
 'negative',
 'negative',
 'negative',
 'negative',
 'negative',
 'negative',
 'negative',
 'negative',
 'negative',
 'neutral',
 'neutral',
 'neutral',
 'neutral',
 'neutral',
 'neutral',
 'neutral',
 'neutral',
 'neutral',
 'neutral']
```

Now let us split the data into train and test.

Function for training data:

```
# to notice: indexing starts at 0
def training_data(data):
    subset_1 = data[:8]  # First 8 elements from 1-10
    subset_2 = data[10:18]  # First 8 elements from 11-20
    subset_3 = data[20:28]  # First 8 elements from rest of the data

    combined_data = subset_1 + subset_2 + subset_3
    return combined_data
```

Function for test data:

```
# to notice: indexing starts at 0
def testing_data(data):
    subset_1 = data[8:10]  # Last 2 elements from 1-10
    subset_2 = data[18:20]  # Last 2 elements from 11-20
    subset_3 = data[28:30]  # Last 2 elements from rest of the data

    combined_data = subset_1 + subset_2 + subset_3
    return combined_data
```

Now, let's use X and y variables as a parameter for the training_data function:

```
X_train = training_data(X)
print(len(X_train))
X_train
```

Output:

```
24
["I was absolutely blown away by the performances in 'Summer's End'. The acting was top-notch, and the plot had me gripped from start to finish. A truly captivating cinematic experience that I would highly recommend.",
 "The special effects in 'Star Battles: Nebula Conflict' were out of this world. I felt like I was actually in space. The storyline was incredibly engaging and left me wanting more. Excellent film.",
 "'The Lost Symphony' was a masterclass in character development and storytelling. The score was hauntingly beautiful and complimented the intense, emotional scenes perfectly. Kudos to the director and cast for creating such a masterpiece.",
 "I was pleasantly surprised by 'Love in the Time of Cholera'. The romantic storyline was heartwarming and the characters were incredibly realistic. The cinematography was also top-notch. A must-watch for all romance lovers.",
 "I went into 'Marble Street' with low expectations, but I was pleasantly surprised. The suspense was well-maintained throughout, and the twist at the end was something I did not see coming. Bravo!",
 "'The Great Plains' is a touching portrayal of life in rural America. The performances were heartfelt and the scenery was breathtaking. I was moved to tears by the end. It's a story that will stay with me for a long time.",
 "The screenwriting in 'Under the Willow Tree' was superb. The dialogue felt real and the characters were well-rounded. The performances were also fantastic. I haven't enjoyed a movie this much in a while.",
 "'Nightshade' is a brilliant take on the superhero genre. The protagonist was relatable and the villain was genuinely scary. The action sequences were thrilling and the storyline was engaging. I can't wait for the sequel.",
 "I was thoroughly disappointed with 'Silver Shadows'. The plot was confusing and the performances were lackluster. I wouldn't recommend wasting your time or this one.",
 "'The Darkened Path' was a disaster. The storyline was unoriginal and the acting was wooden and the special effects were laughably bad. Save your money and skip this one.",
 "I had high hopes for 'The Final Frontier', but it failed to deliver. The plot was full of holes and the characters were poorly developed. It was a disappointing experience.",
 "'The Fall of the Phoenix' was a letdown. The storyline was confusing and the characters were one-dimensional. I found myself checking my watch multiple times throughout the movie.",
 "I regret wasting my time on 'Emerald City'. The plot was nonsensical and the performances were uninspired. It was a major disappointment.",
 "I found 'Hollow Echoes' to be a complete mess. The plot was non-existent, the performances were overdone, and the pacing was all over the place. Definitely not worth the hype.",
 "'Underneath the Stars' was a huge disappointment. The storyline was predictable and the acting was mediocre at best. I was expecting so much more.",
 "I was left unimpressed by 'River's Edge'. The plot was convoluted, the characters were uninteresting, and the ending was unsatisfying. It's a pass for me.",
 "'Remember the Days' was utterly forgettable. The storyline was dull, the performances were bland, and the dialogue was cringeworthy. A big disappointment.",
 "'The Last Frontier' was simply okay. The plot was decent and the performances were acceptable. However, it lacked a certain spark to make it truly memorable.",
 "'Through the Storm' was not bad, but it wasn't great either. The storyline was somewhat predictable, and the characters were somewhat stereotypical. It was an average movie at best.",
 "I found 'After the Rain' to be pretty average. The plot was okay and the performances were decent, but it didn't leave a lasting impression on me.",
 "'Beyond the Horizon' was neither good nor bad. The plot was interesting enough, but the characters were not very well developed. It was an okay watch.",
 "'The Silent Echo' was a mediocre movie. The storyline was passable and the performances were fair, but it didn't stand out in any way.",
 "I thought 'The Scent of Roses' was pretty average. The plot was somewhat engaging, and the performances were okay, but it didn't live up to my expectations.",
 "'Under the Same Sky' was an okay movie. The plot was decent, and the performances were fine, but it lacked depth and originality. It's not a movie I would watch again."]
```

```
y_train = training_data(y)
print(len(y_train))
y_train
```

Output:

```
24
['positive',
 'positive',
 'positive',
 'positive',
 'positive',
 'positive',
 'positive',
 'positive',
 'negative',
 'negative',
 'negative',
 'negative',
 'negative',
 'negative',
 'negative',
 'negative',
 'neutral',
 'neutral',
 'neutral',
 'neutral',
 'neutral',
 'neutral',
 'neutral',
 'neutral']
```

Now, let use X and y variables as a parameter for the testing_data function:

```
X_test = testing_data(X)
print(len(X_test))
X_test
```

Output:

```
6
["The cinematography in 'Awakening' was nothing short of spectacular. The visuals alone are worth the ticket price. The storyline was unique and the performances were solid. An overall fantastic film ",
 "'Eternal Embers' was a cinematic delight. The storytelling was original and the performances were exceptional. The director's vision was truly brought to life on the big screen. A must-see for all movie lovers.",
 "The acting in 'Desert Mirage' was subpar, and the plot was boring. I found myself yawning multiple times throughout the movie. Save your time and skip this one.",
 "'Crimson Dawn' was a major letdown. The plot was cliched and the characters were flat. The special effects were also poorly executed. I wouldn't recommend it.",
 "'Chasing Shadows' was fairly average. The plot was not bad, and the performances were passable, but it lacked a certain spark. It was just okay.",
 "'Beneath the Surface' was pretty run-of-the-mill. The plot was decent, the performances were okay, but it wasn't particularly memorable. It was an okay movie."]
```

```
y_test = testing_data(y)
print(len(y_test))
y_test
```

Output:

```
6
['positive', 'positive', 'negative', 'negative', 'neutral', 'neutral']
```

Defining and training the OpenAI model:

```
# defining the openai model to use
clf = ZeroShotGPTClassifier(openai_model="gpt-3.5-turbo")

# fitting the data
clf.fit(X_train, y_train)
```

```
▾ ZeroShotGPTClassifier
ZeroShotGPTClassifier()
```

Predict on X_test using the clf model:

```
%%time
# predicting the data
predicted_labels = clf.predict(X_test)
```

Output:

```
 17%|█▋        | 1/6 [00:09<00:47,  9.44s/it]Could not obtain the completion after 3 retries: `RateLimitError :: You exceeded your current quota, please check your plan and billing details.`
None
Could not extract the label from the completion: 'NoneType' object is not subscriptable
 33%|███▎      | 2/6 [00:18<00:37,  9.30s/it]Could not obtain the completion after 3 retries: `RateLimitError :: You exceeded your current quota, please check your plan and billing details.`
None
Could not extract the label from the completion: 'NoneType' object is not subscriptable
 50%|█████     | 3/6 [00:29<00:29,  9.84s/it]Could not obtain the completion after 3 retries: `RateLimitError :: You exceeded your current quota, please check your plan and billing details.`
None
Could not extract the label from the completion: 'NoneType' object is not subscriptable
 67%|██████▋   | 4/6 [00:38<00:19,  9.58s/it]Could not obtain the completion after 3 retries: `RateLimitError :: You exceeded your current quota, please check your plan and billing details.`
None
Could not extract the label from the completion: 'NoneType' object is not subscriptable
 83%|████████▎ | 5/6 [00:47<00:09,  9.47s/it]Could not obtain the completion after 3 retries: `RateLimitError :: You exceeded your current quota, please check your plan and billing details.`
None
Could not extract the label from the completion: 'NoneType' object is not subscriptable
100%|██████████| 6/6 [00:56<00:00,  9.46s/it]Could not obtain the completion after 3 retries: `RateLimitError :: You exceeded your current quota, please check your plan and billing details.`
None
Could not extract the label from the completion: 'NoneType' object is not subscriptable
CPU times: user 455 ms, sys: 41.8 ms, total: 497 ms
Wall time: 56.8 s
```

Tagging the predictions for each sentence:

```
for review, sentiment in zip(X_test, predicted_labels):
    print(f"Review: {review}\nPredicted Sentiment: {sentiment}\n\n")
```

Output:

```
Review: The cinematography in 'Awakening' was nothing short of spectacular. The visuals alone are worth the ticket price. The storyline was unique and the performances were solid. An overall fantasti
Predicted Sentiment: positive

Review: 'Eternal Embers' was a cinematic delight. The storytelling was original and the performances were exceptional. The director's vision was truly brought to life on the big screen. A must-see fo
Predicted Sentiment: negative

Review: The acting in 'Desert Mirage' was subpar, and the plot was boring. I found myself yawning multiple times throughout the movie. Save your time and skip this one.
Predicted Sentiment: neutral

Review: 'Crimson Dawn' was a major letdown. The plot was cliched and the characters were flat. The special effects were also poorly executed. I wouldn't recommend it.
Predicted Sentiment: positive

Review: 'Chasing Shadows' was fairly average. The plot was not bad, and the performances were passable, but it lacked a certain spark. It was just okay.
Predicted Sentiment: negative

Review: 'Beneath the Surface' was pretty run-of-the-mill. The plot was decent, the performances were okay, but it wasn't particularly memorable. It was an okay movie.
Predicted Sentiment: positive
```

Evaluate model:

```
from sklearn.metrics import accuracy_score
print(f"Accuracy: {accuracy_score(y_test, predicted_labels):.2f}")
```

Output:

```
Accuracy: 1.00
```

Scikit-LLM goes the extra mile by ensuring that the responses it receives contain valid labels. When it encounters a response without a valid label, Scikit-LLM doesn't leave you hanging. Instead, it steps in and selects a label randomly, taking into account the probabilities based on how frequently those labels appeared in the training data.

To put it simply, Scikit-LLM takes care of the technical details, making sure you always have meaningful labels to work with. It's got your back, even if a response happens to be missing a label, as it will intelligently choose one for you based on its knowledge of label frequencies in the training data.

What If You Find Yourself Without Labeled Data?

Here's the intriguing aspect: you don't actually require prelabeled data to train the model. Instead, all you need is a list of potential candidate labels to get started. This approach opens up possibilities for training models even when you don't have the luxury of pre-existing labeled datasets.

Defining the training OpenAI model:

```
# defining the model
clf_no_label = ZeroShotGPTClassifier()
```

```
# No training so passing the labels only for prediction
clf_no_label.fit(None, ['positive', 'negative', 'neutral'])
```

Predict on X_test using the model:

```
# predicting the labels
predicted_labels_without_training_data = clf_no_label.predict(X_test)
predicted_labels_without_training_data
```

Output:

```
 17%|█        | 1/6 [00:10<00:54, 10.80s/it]Could not obtain the completion after 3 retries: `RateLimitError :: You exceeded your current quota, please check your plan and billing details.`
None
Could not extract the label from the completion: 'NoneType' object is not subscriptable
 33%|██       | 2/6 [00:20<00:39,  9.94s/it]Could not obtain the completion after 3 retries: `RateLimitError :: You exceeded your current quota, please check your plan and billing details.`
None
Could not extract the label from the completion: 'NoneType' object is not subscriptable
 50%|████     | 3/6 [00:29<00:20,  9.67s/it]Could not obtain the completion after 3 retries: `RateLimitError :: You exceeded your current quota, please check your plan and billing details.`
None
Could not extract the label from the completion: 'NoneType' object is not subscriptable
 67%|██████   | 4/6 [00:38<00:18,  9.50s/it]Could not obtain the completion after 3 retries: `RateLimitError :: You exceeded your current quota, please check your plan and billing details.`
None
Could not extract the label from the completion: 'NoneType' object is not subscriptable
 83%|███████  | 5/6 [00:47<00:09,  9.39s/it]Could not obtain the completion after 3 retries: `RateLimitError :: You exceeded your current quota, please check your plan and billing details.`
None
Could not extract the label from the completion: 'NoneType' object is not subscriptable
100%|█████████| 6/6 [00:57<00:00,  9.52s/it]Could not obtain the completion after 3 retries: `RateLimitError :: You exceeded your current quota, please check your plan and billing details.`
None
Could not extract the label from the completion: 'NoneType' object is not subscriptable

['neutral', 'neutral', 'negative', 'positive', 'positive', 'positive']
```

Tagging the predictions for each sentence:

```
for review, sentiment in zip(X_test, predicted_labels_without_
training_data):
    print(f"Review: {review}\nPredicted Sentiment: {sentiment}\n\n")
```

Output:

```
Review: The cinematography in 'Awakening' was nothing short of spectacular. The visuals alone are worth the ticket price. The storyline was unique and the performances were solid. An overall fantas
Predicted Sentiment: neutral

Review: 'Eternal Embers' was a cinematic delight. The storytelling was original and the performances were exceptional. The director's vision was truly brought to life on the big screen. A must-see
Predicted Sentiment: neutral

Review: The acting in 'Desert Mirage' was subpar, and the plot was boring. I found myself yawning multiple times throughout the movie. Save your time and skip this one.
Predicted Sentiment: negative

Review: 'Crimson Dawn' was a major letdown. The plot was cliched and the characters were flat. The special effects were also poorly executed. I wouldn't recommend it.
Predicted Sentiment: positive

Review: 'Chasing Shadows' was fairly average. The plot was not bad, and the performances were passable, but it lacked a certain spark. It was just okay.
Predicted Sentiment: positive

Review: 'Beneath the Surface' was pretty run-of-the-mill. The plot was decent, the performances were okay, but it wasn't particularly memorable. It was an okay movie.
Predicted Sentiment: positive
```

Evaluate model:

```
print(f"Accuracy: {accuracy_score(y_test, predicted_labels_without_
training_data):.2f}")
```

Output:

```
Accuracy: 1.00
```

Till now we explored how to use Scikit-LLM models for text classification, next we will explore the other features of Scikit-LLM.

Note: In the next examples, we will not split the data into train and test or evaluate the model like we did for text classification instead focus on the usage part.

Multilabel Zero-Shot Text Classification

Conducting multilabel zero-shot text classification might sound complex, but it's actually more straightforward than you'd imagine.

Implementation

```
# importing Multi-Label zeroshot module and classification dataset
from skllm import MultiLabelZeroShotGPTClassifier
from skllm.datasets import get_multilabel_classification_dataset
# get classification dataset from sklearn
X, y = get_multilabel_classification_dataset()

# defining the model
clf = MultiLabelZeroShotGPTClassifier(max_labels=3)

# fitting the model
clf.fit(X, y)

# making predictions
labels = clf.predict(X)
```

The only distinction between zero-shot and multilabel zero-shot classification lies in the creation of an instance of the MultiLabelZeroShotGPTClassifier class. In the case of multilabel zero-shot classification, you specify the maximum number of labels you want to assign to each sample, like setting max_labels=3 as an example. This parameter allows you to control how many labels the model can assign to a given text sample during classification.

What If You Find Yourself Without Labeled Data?

In the scenario outlined earlier, the MultiLabelZeroShotGPTClassifier can still be trained effectively. Instead of using traditional labeled data (X and y), you can train the classifier by providing a list of potential candidate labels. In this setup, the "y" component should be structured as a List of Lists, where each inner list contains candidate labels for a specific text sample.

Here's an example illustrating the training process without labeled data:

Implementation

```
# getting classification dataset for prediction only
from skllm.datasets import get_multilabel_classification_dataset
from skllm import MultiLabelZeroShotGPTClassifier
X, _ = get_multilabel_classification_dataset()
# Defining all the labels that need to be predicted
candidate_labels = [
    "Quality",
    "Price",
    "Delivery",
    "Service",
    "Product Variety"
]

# creating the model
clf = MultiLabelZeroShotGPTClassifier(max_labels=3)

# fitting the labels only
clf.fit(None, [candidate_labels])

# predicting the data
labels = clf.predict(X)
```

Text Vectorization

Text vectorization is a crucial process that involves transforming textual information into numerical format, enabling machines to comprehend and analyze it effectively. Within the Scikit-LLM framework, you'll find a valuable tool called the GPTVectorizer. This module serves the purpose of converting text, regardless of its length, into a fixed-size set of numerical values known as a vector. This transformation allows machine learning models to process and make sense of text-based data more efficiently.

Implementation

```
# Importing the GPTVectorizer class from the skllm.preprocessing module
from skllm.preprocessing import GPTVectorizer

# Creating an instance of the GPTVectorizer class and assigning it to the
variable 'model'
model = GPTVectorizer()

# transforming the
vectors = model.fit_transform(X)
```

When you apply the "fit_transform" method of the GPTVectorizer instance to your input data "X," it not only fits the model to the data but also transforms the text into fixed-dimensional vectors. These resulting vectors are then stored in a variable, conventionally named "vectors."

Let's illustrate an example of how to integrate the GPTVectorizer with the XGBoost Classifier in a scikit-learn pipeline. This approach enables you to efficiently preprocess text and perform classification tasks seamlessly:

```
# Importing the necessary modules and classes
from sklearn.pipeline import Pipeline
from sklearn.preprocessing import LabelEncoder
from xgboost import XGBClassifier

# Creating an instance of LabelEncoder class
le = LabelEncoder()

# Encoding the training labels 'y_train' using LabelEncoder
y_train_encoded = le.fit_transform(y_train)
```

```
# Encoding the test labels 'y_test' using LabelEncoder
y_test_encoded = le.transform(y_test)

# Defining the steps of the pipeline as a list of tuples
steps = [('GPT', GPTVectorizer()), ('Clf', XGBClassifier())]

# Creating a pipeline with the defined steps
clf = Pipeline(steps)

# Fitting the pipeline on the training data 'X_train' and the encoded
training labels 'y_train_encoded'
clf.fit(X_train, y_train_encoded)

# Predicting the labels for the test data 'X_test' using the trained
pipeline
yh = clf.predict(X_test)
```

Text Summarization

Indeed, GPT excels at text summarization, and this strength is harnessed within
Scikit-LLM through the GPTSummarizer module. You can utilize this module in two
distinct ways:

1. Standalone Summarization: You can use GPTSummarizer on
   its own to generate concise and coherent summaries of textual
   content, making it easier to grasp the main points of lengthy
   documents.

2. As a Preprocessing Step: Alternatively, you can integrate
   GPTSummarizer into a broader workflow as a preliminary step
   before performing other operations. For example, you can use it to
   reduce the size of text data while retaining essential information.
   This enables more efficient handling of text-based data without
   compromising the content's quality and relevance.

Implementation

```
# Importing the GPTSummarizer class from the skllm.preprocessing module
from skllm.preprocessing import GPTSummarizer

# Importing the get_summarization_dataset function
from skllm.datasets import get_summarization_dataset

# Calling the get_summarization_dataset function
X = get_summarization_dataset()

# Creating an instance of the GPTSummarizer
s = GPTSummarizer(openai_model='gpt-3.5-turbo', max_words=15)

# Applying the fit_transform method of the GPTSummarizer instance to the
input data 'X'.
# It fits the model to the data and generates the summaries, which are
assigned to the variable 'summaries'
summaries = s.fit_transform(X)
```

It's important to understand that the "max_words" hyperparameter serves as a flexible guideline for limiting the number of words in the generated summaries. It's not strictly enforced beyond the initial prompt you provide. In practical terms, this means that there might be instances where the actual number of words in the generated summaries slightly exceeds the specified limit.

In simpler terms, while "max_words" provides an approximate target for the summary length, the summarizer may occasionally produce slightly longer summaries. This behavior depends on the specific context and content of the input text as the summarizer aims to maintain coherence and relevance in its output.

Conclusion

Basically, Scikit-LLM can be used for text analysis, and it is designed to be easy to use and provides a range of features, including zero-shot text classification, multilabel zero-shot text classification, text vectorization, text translation, and text summarization.

The most important thing is that you don't require prelabeled data to train any models. That's the beauty of Scikit-LLMs.

To get started with using LLMs for text analysis easily. Scikit-LLM provides a simple and intuitive API that makes it easy to get started with using LLMs for text analysis, even if you are not familiar with LLMs or machine learning.

To combine LLMs with other machine learning algorithms. Scikit-LLM can be integrated with scikit-learn pipelines, which makes it easy to combine LLMs with other machine learning algorithms. This can be useful for complex text analysis tasks that require multiple steps.

To experiment with LLMs for text analysis. Scikit-LLM is an open source project, which means that it is free to use and modify. This makes it a good option for researchers and developers who want to experiment with LLMs for text analysis:

- You can use Scikit-LLM to classify customer feedback into different categories, such as positive, negative, or neutral. This information can be used to improve customer service or product development.

- You can use Scikit-LLM to classify news articles into different topics, such as politics, business, or sports. This information can be used to create personalized news feeds or to track trends in the news.

- You can use Scikit-LLM to translate documents from one language to another. This can be useful for businesses that operate in multiple countries or for people who want to read documents in a language they don't speak.

- You can use Scikit-LLM to summarize long text documents. This can be useful for quickly getting the main points of a document or for creating a shorter version of a document for publication.

In addition to the mentioned earlier, Scikit-LLM also offers a number of other benefits, such as

- Accuracy: Scikit-LLM has been shown to be accurate in a number of text analysis tasks, including zero-shot text classification and text summarization.

- Speed: Scikit-LLM is relatively fast, which makes it suitable for tasks that require real-time processing.

Scalability: Scikit-LLM can be scaled to handle large amounts of text data.

CHAPTER 7

LLMs for Enterprise and LLMOps

In this chapter, we are presenting a reference framework for the emerging app stack of large language models (LLMs). The framework illustrates the prevalent systems, tools, and design approaches that have been observed in practice among AI startups and enterprises. It's important to note that this stack is in its nascent stages and is likely to undergo significant transformations with the progression of underlying technology. Nevertheless, our intention is for this resource to provide valuable guidance to developers who are presently engaged with LLMs.

Numerous approaches exist for harnessing the capabilities of LLMs in development, which encompass creating models from scratch, refining open source models through fine-tuning, or utilizing hosted APIs. The framework we're presenting here is centered around in-context learning, a prevalent design strategy that most developers opt for, particularly made feasible through foundational models.

The subsequent section offers a succinct elucidation of this strategy, with experienced LLM developers having the option to skip it.

The power of LLMs lies not only in their capabilities but also in their responsible and ethical usage, which is paramount in enterprise settings. We'll discuss how organizations are navigating the intricate landscape of data privacy, bias mitigation, and transparency while harnessing the transformative potential of these language models.

Now, as we prepare to conclude our exploration, it's important to highlight a crucial enabler of this transformation: cloud services. The cloud, with its unparalleled computational power, scalability, and global reach, has become the infrastructure of choice for deploying and managing LLMs. It provides a dynamic environment where businesses can harness the full potential of these language models while enjoying a range of benefits. We'll briefly touch upon how cloud services complement the adoption of LLMs, offering scalability, cost-efficiency, security, and seamless integration with existing workflows. Here are three ways you can enable LLMs at enterprise.

© Akshay Kulkarni, Adarsha Shivananda, Anoosh Kulkarni, Dilip Gudivada 2023
A. Kulkarni et al., *Applied Generative AI for Beginners*, https://doi.org/10.1007/978-1-4842-9994-4_7

Private Generalized LLM API

A private generalized LLM API is a way for enterprises to access a large language model (LLM) that has been trained on a massive dataset of text and code. The API is private, which means that the enterprise is the only one who can use it. This ensures that the enterprise's data is kept private.

There are several benefits to using a private generalized LLM API:

- First, it allows enterprises to customize the LLM to their specific needs. For example, the enterprise can specify the LLM's training data, the LLM's architecture, and the LLM's parameters. This allows the enterprise to get the most out of the LLM for their specific tasks.

- Second, a private generalized LLM API is more secure than using a public LLM API. This is because the enterprise's data is not shared with anyone else. This is important for enterprises that are concerned about the security of their data.

- Third, a private generalized LLM API is more scalable than using a public LLM API. This is because the enterprise can increase the amount of computing power that is used to train and run the LLM. This allows the enterprise to use the LLM for more demanding tasks.

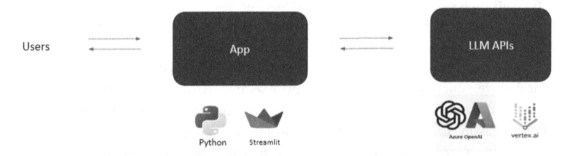

Figure 7-1. *Private generalized LLM API*

However, there are also some challenges to using a private generalized LLM API:

- It can be expensive to develop and maintain a private LLM API. This is because the enterprise needs to have the expertise and resources to train and run the LLM.

- A private LLM API can be slower than using a public LLM API. This is because the enterprise's data needs to be transferred to the LLM before it can be processed.

- A private LLM API can be less flexible than using a public LLM API. This is because the enterprise is limited to the features and capabilities that are provided by the API.

Overall, a private generalized LLM API is a good option for enterprises that need to use an LLM for their specific tasks and that are concerned about the security of their data. However, it is important to weigh the benefits and challenges of using a private LLM API before making a decision.

Here are some examples of how enterprises can use a private generalized LLM API:

- Customer Service: An enterprise can use an LLM to generate personalized responses to customer queries.

- Product Development: An enterprise can use an LLM to generate ideas for new products and services.

- Marketing: An enterprise can use an LLM to create personalized marketing campaigns.

- Risk Management: An enterprise can use an LLM to identify potential risks and vulnerabilities.

- Fraud Detection: An enterprise can use an LLM to detect fraudulent transactions.

Design Strategy to Enable LLMs for Enterprise: In-Context Learning

At its core, in-context learning involves employing off-the-shelf LLMs (without fine-tuning) and manipulating their behavior via astute prompts and conditioning based on private "contextual" data.

Consider the scenario of crafting a chatbot to address queries related to a collection of legal documents. A straightforward approach might involve inserting all documents into a ChatGPT or GPT-4 prompt, followed by posing questions about them. While this might suffice for minute datasets, it isn't scalable. The largest GPT-4 model can only handle around 50 pages of input text, and its performance in terms of inference time and accuracy degrades significantly as this context window limit is approached.

In-context learning tackles this quandary ingeniously by adopting a stratagem: instead of supplying all documents with each LLM prompt, it dispatches only a select set of the most pertinent documents. These pertinent documents are determined with the aid of—you guessed it—LLMs.

In broad strokes, the workflow can be partitioned into three phases:

> Data Preprocessing/Embedding: This phase entails storing private data (e.g., legal documents) for future retrieval. Typically, the documents are divided into sections, processed through an embedding model, and subsequently stored in a specialized database called a vector database.

> Prompt Construction/Retrieval: Upon a user submitting a query (such as a legal question), the application generates a sequence of prompts for the language model. A compiled prompt usually amalgamates a developer-defined prompt template, instances of valid outputs known as few-shot examples, any requisite data retrieved from external APIs, and a selection of pertinent documents obtained from the vector database.

> Prompt Execution/Inference: Once the prompts are compiled, they are fed into a pre-trained LLM for inference, encompassing both proprietary model APIs and open source or self-trained models. In some instances, developers supplement operational systems like logging, caching, and validation during this phase.

Though this may appear intricate, it's often simpler than the alternative: training or fine-tuning the LLM itself. In-context learning doesn't necessitate a dedicated team of machine learning engineers. Additionally, you're not compelled to manage your own infrastructure or invest in costly dedicated instances from OpenAI. This approach essentially transforms an AI challenge into a data engineering task, a domain that many startups and established companies are already familiar with. It generally surpasses fine-tuning for moderately small datasets—given that specific information needs to be present in the training set multiple times for an LLM to retain it via fine-tuning—and it can swiftly incorporate new data in almost real time.

A pivotal query about in-context learning pertains to altering the underlying model to expand the context window. This is indeed a possibility and is an active area of research. Nonetheless, this introduces a range of trade-offs, primarily the quadratic escalation of inference costs and time with the extension of prompt length. Even linear expansion (the most favorable theoretical outcome) would prove cost-prohibitive for many applications today. Presently, executing a single GPT-4 query over 10,000 pages would translate to hundreds of dollars based on prevailing API rates.

Context Injection Architecture

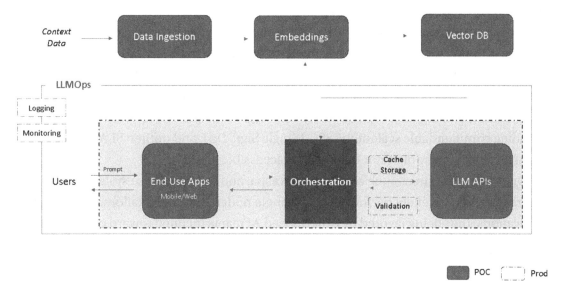

Figure 7-2. *Context injection architecture*

Data Preprocessing/Embedding

Contextual data for LLM applications encompasses various formats, including text documents, PDFs, and structured data like CSV or SQL tables. The methods for loading and transforming this data exhibit considerable diversity among the developers we've engaged with. Many opt for conventional ETL tools like Databricks or Airflow. A subset also utilizes document loaders integrated into orchestration frameworks such as LangChain (powered by Unstructured) and LlamaIndex (powered by Llama Hub). Nevertheless, we perceive this aspect of the framework to be relatively underdeveloped, thereby presenting an opportunity for purpose-built data replication solutions tailored to LLM applications.

In the realm of embeddings, the majority of developers make use of the OpenAI API, particularly the text-embedding-ada-002 model. This model is user-friendly, especially for those already acquainted with other OpenAI APIs, yielding reasonably satisfactory outcomes and progressively more cost-effective. In certain contexts, larger enterprises are also exploring Cohere, a platform that specializes more narrowly in embeddings and exhibits superior performance under specific scenarios. For developers inclined toward open source options, the Hugging Face Sentence Transformers library stands as a standard choice. Furthermore, the potential exists to generate distinct types of embeddings customized to varying use cases—an aspect that presently represents a niche practice but holds promise as a realm of research.

From a system's perspective, the pivotal component within the preprocessing pipeline is the vector database. Its role involves the efficient storage, comparison, and retrieval of countless embeddings (or vectors). Pinecone emerges as the most prevalent selection in the market, primarily due to its cloud-hosted nature, facilitating easy initiation and offering an array of features that larger enterprises require for production, including commendable scalability, SSO (Single Sign-On), and uptime SLAs.

An extensive array of vector databases is accessible, however:

Open Source Systems such as Weaviate, Vespa, and Qdrant: These systems generally exhibit excellent performance on single nodes and can be tailored for specific applications, thus being favored by experienced AI teams inclined toward constructing bespoke platforms.

Local Vector Management Libraries like Chroma and Faiss: These offer a positive developer experience and can be rapidly set up for smaller applications and development experiments. However, they may not completely replace a comprehensive database at larger scales.

OLTP Extensions like Pgvector: This is a suitable option for developers who attempt to integrate Postgres for every database requirement or enterprises that predominantly source their data infrastructure from a single cloud provider. Nonetheless, the long-term integration of vector and scalar workloads remains unclear.

In terms of future prospects, many open source vector database providers are venturing into cloud offerings. Our research suggests that achieving robust cloud performance across a diverse landscape of potential use cases is a formidable challenge. Consequently, while the array of options may not witness substantial immediate changes, long-term shifts are likely. The pivotal question revolves around whether vector databases will parallel their OLTP and OLAP counterparts by converging around one or two widely embraced systems.

Another unresolved query pertains to how embeddings and vector databases will evolve alongside the expansion of the usable context window for most models. It might seem intuitive to assume that embeddings will become less essential as contextual data can be directly integrated into prompts. Contrarily, insights from experts in this domain suggest the opposite—that the significance of the embedding pipeline might intensify over time. Although extensive context windows offer considerable utility, they also entail notable computational costs, thereby necessitating efficient utilization. We might witness a surge in popularity for diverse types of embedding models, trained explicitly for model relevance, coupled with vector databases crafted to facilitate and capitalize on these advancements.

Prompt Construction/Retrieval

Interacting with large language models (LLMs) involves a structured process that resembles a generalized API call. Developers create requests in the form of prompt templates, submit them to the model, and subsequently parse the output to ensure correctness and relevance. This interaction process has become increasingly sophisticated, allowing developers to integrate contextual data and orchestrate nuanced responses, which is crucial for various applications.

Approaches for eliciting responses from LLMs and integrating contextual data are progressively growing in complexity and significance, emerging as a pivotal avenue for distinguishing products. During the inception of new projects, most developers commence with experimentation involving uncomplicated prompts. These prompts might entail explicit directives (zero-shot prompts) or even instances of expected outputs (few-shot prompts). While such prompts often yield favorable outcomes, they tend to fall short of the accuracy thresholds necessary for actual production deployments.

The subsequent tier of prompting strategy, often referred to as "prompting jiu-jitsu," is geared toward anchoring model responses in some form of verifiable information and introducing external context that the model hasn't been exposed to during training. The Prompt Engineering Guide delineates no less than 12 advanced prompting strategies, which include chain-of-thought, self-consistency, generated knowledge, tree of thoughts, directional stimulus, and several others. These strategies can also be synergistically employed to cater to diverse LLM applications, spanning from document-based question answering to chatbots, and beyond.

This is precisely where orchestration frameworks like LangChain and LlamaIndex prove their mettle. These frameworks abstract numerous intricacies associated with prompt chaining, interfacing with external APIs (including discerning when an API call is warranted), retrieving contextual data from vector databases, and maintaining coherence across multiple LLM interactions. Additionally, they furnish templates tailored to numerous commonly encountered applications. The output they provide takes the form of a prompt or a sequence of prompts to be submitted to a language model. These frameworks are widely embraced by hobbyists and startups striving to kick-start their applications, with LangChain reigning as the front-runner.

While LangChain is a relatively recent endeavor (currently at version 0.0.201), instances of applications constructed with it are already transitioning into the production phase. Some developers, particularly those who embraced LLMs in their early stages, might opt to switch to raw Python in production to circumvent additional dependencies. However, we anticipate this do-it-yourself approach to dwindle over time across the majority of use cases, much akin to the evolution observed in the traditional web app stack.

In the current landscape, OpenAI stands at the forefront of language models. Nearly all developers we've interacted with initiate new LLM applications using the OpenAI API, predominantly opting for models such as gpt-4 or gpt-4-32k. This choice offers an optimal scenario for application performance, boasting ease of use across a diverse spectrum of input domains, typically necessitating no fine-tuning or self-hosting.

As projects progress into the production phase and aim for scalability, a wider array of choices emerges. Several common approaches we encountered include the following:

Transitioning to gpt-3.5-turbo: This option stands out due to its approximately 50-fold cost reduction and significantly enhanced speed compared to GPT-4. Many applications don't require the precision levels of GPT-4 but do demand low-latency inference and cost-effective support for free users.

Exploring Other Proprietary Vendors (Particularly Anthropic's Claude Models): Claude models provide rapid inference, accuracy akin to GPT-3.5, greater customization flexibility for substantial clientele, and the potential to accommodate a context window of up to 100k (though we've observed accuracy decline with longer inputs).

Prioritizing Certain Requests for Open Source Models: This tactic can be especially effective for high-volume B2C scenarios like search or chat, where query complexity varies widely and there's a need to serve free users economically. This approach often pairs well with fine-tuning open source base models. While we don't delve deeply into the specifics of this tooling stack in this article, platforms such as Databricks, Anyscale, Mosaic, Modal, and RunPod are increasingly adopted by numerous engineering teams.

Diverse inference options exist for open source models, ranging from straightforward API interfaces provided by Hugging Face and Replicate to raw computational resources from major cloud providers, and more opinionated cloud offerings like those mentioned earlier.

Presently, open source models lag behind their proprietary counterparts, yet the gap is narrowing. Meta's LLaMa models have established a new benchmark for open source accuracy, sparking a proliferation of variations. Since LLaMa's licensing restricts it to research use only, various new providers have stepped in to develop alternative base models (examples include Together, Mosaic, Falcon, and Mistral). Meta is also contemplating a potentially fully open source release of LLaMa 2.

Anticipating the eventuality when open source LLMs achieve accuracy levels on par with GPT-3.5, we foresee a moment akin to Stable Diffusion for text, marked by extensive experimentation, sharing, and operationalization of fine-tuned models. Hosting companies like Replicate are already incorporating tooling to facilitate developers' consumption of these models. There's an increasing belief among developers that smaller, fine-tuned models can attain cutting-edge precision in specific use cases.

A majority of developers we engaged with haven't delved deeply into operational tooling for LLMs at this juncture. Caching, typically built on Redis, is relatively widespread as it enhances application response times while being cost-effective. Tools like Weights & Biases and MLflow (adapted from traditional machine learning) or LLM-focused solutions like PromptLayer and Helicone are also commonly utilized. These tools enable logging, tracking, and evaluation of LLM outputs, often for purposes such as enhancing prompt construction, refining pipelines, or model selection. Additionally, several new tools are in development to validate LLM outputs (e.g., Guardrails) or identify prompt injection attacks (e.g., Rebuff). Most of these operational tools encourage the use of their own Python clients to initiate LLM calls, prompting curiosity regarding how these solutions will coexist over time.

Fine-Tuning

Fine-tuning with transfer learning is a technique that uses a pre-trained LLM as a starting point for training a new model on a specific task or domain. This can be done by freezing some of the layers of the pre-trained LLM and only training the remaining layers. This helps to prevent the model from overfitting to the new data and ensures that it still retains the general knowledge that it learned from the pre-trained LLM.

The following are the steps involved in fine-tuning with transfer learning:

1. Choose a Pre-trained LLM: There are many different LLMs available, each with its own strengths and weaknesses. The choice of LLM will depend on the specific task or domain that you want to fine-tune the model for.

2. Collect a Dataset of Text and Code That Is Specific to the Task or Domain: The size and quality of the dataset will have a significant impact on the performance of the fine-tuned model.

3. Prepare the Dataset for Fine-Tuning: This may involve cleaning the data, removing duplicate entries, and splitting the data into training and test sets.

4. Freeze Some of the Layers of the Pre-trained LLM: This can be done by setting the learning rate of the frozen layers to zero.

5. Train the Remaining Layers of the LLM on the Training Set: This is done by using a supervised learning algorithm to adjust the parameters of the remaining layers so that they can better predict the correct output for the given input.

6. Evaluate the Fine-Tuned Model on the Test Set: This will give you an idea of how well the model has learned to perform the task.

Fine-tuning with transfer learning can be a very effective way to improve the performance of LLMs on a wide variety of tasks. However, it is important to note that the performance of the fine-tuned model will still depend on the quality of the dataset that is used to fine-tune the model. Here is an example of fine-tuning in Figure 7-3.

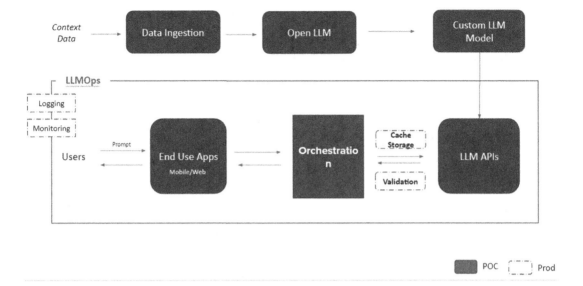

Figure 7-3. *Fine-Tuning*

Here are some of the benefits of fine-tuning with transfer learning:

- It can save time and resources. Transfer learning can be used to fine-tune a model on a new task without having to train the model from scratch.

- It can improve performance. Transfer learning can help to improve the performance of a model on a new task by leveraging the knowledge that the model has already learned from the pre-trained LLM.

- It can make models more generalizable. Transfer learning can help to make models more generalizable to new tasks by reducing the amount of data that is needed to train the model.

However, there are also some challenges associated with fine-tuning with transfer learning:

- It can be difficult to choose the right hyperparameters for the fine-tuning process.

- It can be difficult to find a pre-trained LLM that is a good fit for the new task.

- It can be difficult to prevent the model from overfitting to the new data.

Overall, fine-tuning with transfer learning is a powerful technique that can be used to improve the performance of LLMs on a wide variety of tasks. However, it is important to weigh the benefits and challenges of fine-tuning with transfer learning before making a decision.

Technology Stack
Gen AI/LLM Testbed

To harness the full potential of LLMs and ensure their responsible development, it is crucial to establish a dedicated LLM testbed. This testbed serves as a controlled environment for researching, testing, and evaluating LLMs, facilitating innovation while addressing ethical, safety, and performance concerns. Here is a sample testbed that could be used.

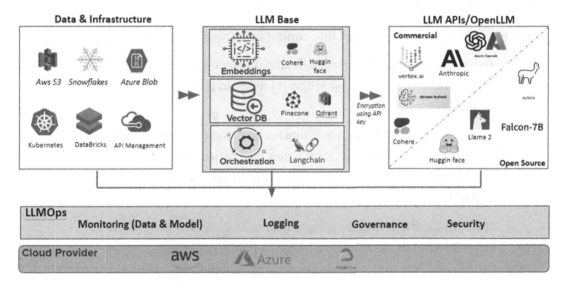

Figure 7-4. *Gen AI/LLM testbed*

Designing a technology stack for generative AI involves selecting and integrating various tools, frameworks, and platforms that facilitate the development, training, and deployment of generative models. Figure 7-5 shows an outline of a technology stack that you might consider.

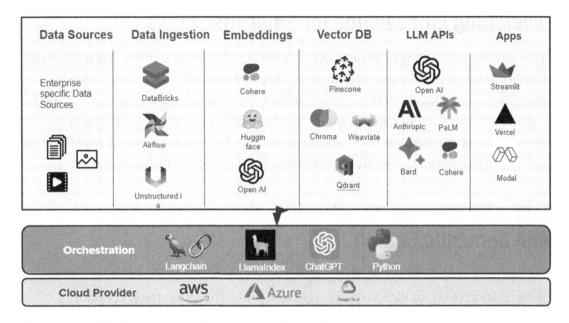

Figure 7-5. *Technology stack for generative AI*

Data Sources

Data sources are a critical component of any generative AI project. The quality, diversity, and quantity of data you use can significantly impact the performance and capabilities of your generative models.

Data Processing

In the journey to enable large language models (LLMs) for enterprise applications, harnessing specialized data processing services is pivotal to efficiently manage the intricacies of data preparation and transformation. While several services contribute to this realm, three stand out as key players: Databricks, Apache Airflow, and tools like Unstructured.io to process unstructured data. It's imperative to acknowledge that alongside these options, a multitude of alternatives also shape the landscape of data processing services.

Leveraging Embeddings for Enterprise LLMs

In the journey of enabling large language models (LLMs) for enterprises, the integration of embeddings serves as a potent strategy to enhance semantic understanding. Embeddings, compact numerical representations of words and documents, are pivotal in enabling LLMs to comprehend context, relationships, and meanings. This section delves into how embeddings from prominent sources like Cohere, OpenAI, and Hugging Face can be harnessed to amplify the effectiveness of LLMs within enterprise contexts.

Vector Databases: Accelerating Enterprise LLMs with Semantic Search

In the pursuit of optimizing large language models (LLMs) for enterprise applications, the integration of vector databases emerges as a game-changing strategy. Vector databases, including solutions like Pinecone, Chroma, Weaviate, and Qdrant, revolutionize the efficiency of semantic search and content retrieval. This subsection delves into how these vector databases can be seamlessly integrated into LLM workflows, thereby enhancing the speed and precision of content retrieval within enterprise contexts.

LLM APIs: Empowering Enterprise Language Capabilities

In the realm of enterprise language capabilities, the utilization of large language model (LLM) APIs has emerged as a cornerstone strategy. These APIs, including offerings from OpenAI, Anthropic, Palm, Bard, and Cohere, grant enterprises seamless access to cutting-edge language processing capabilities. This section delves into how these LLM APIs can be harnessed to elevate communication, content generation, and decision-making within enterprise contexts.

However, you could also use a private generalized LLM Api for your own use case as shown in Figure 7-6.

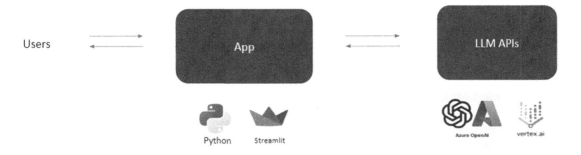

Figure 7-6. *Private generalized LLM API*

LLMOps

What Is LLMOps?

The LLMOps (large language model operations) platform offers a well-defined, comprehensive workflow that covers training, optimization, deployment, and continuous monitoring of LLMs, whether they are open source or proprietary. This streamlined approach is designed to expedite the implementation of generative AI models and their applications.

As organizations increasingly integrate LLMs into their operations, it becomes essential to establish robust and efficient LLMOps. This section delves into the significance of LLMOps and how it ensures the reliability and efficiency of LLMs in enterprise settings.

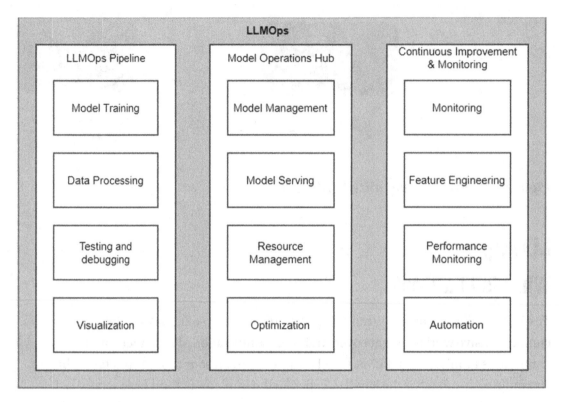

Figure 7-7. *LLMOps*

Sustaining oversight of generative AI models and applications hinges on the ongoing monitoring process, aimed at addressing challenges such as data drift and other factors that may impede their capacity to produce accurate and secure results.

Figure 7-8 represents the LLMOps workflow.

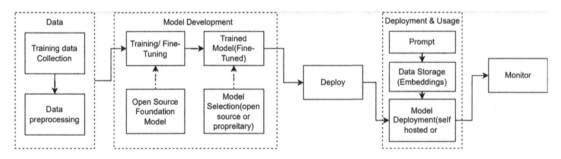

Figure 7-8. *LLMOps workflow*

Why LLMOps?

- Computational Resources: Efficient resource allocation, fine-tuning models, optimizing storage, and managing computational demands, ensuring effective deployment and operation of LLMs becomes key.

- Model Fine-Tuning: Fine-tuning of pre-trained large language models (LLMs) may be necessary to tailor them for specific tasks or datasets, ensuring their optimal performance in practical applications.

- Ethical Concerns: Large language models (LLMs) have the capability to generate content, but ethical concerns arise when they are employed to produce harmful or offensive material.

- Hallucinations: LLM "imagines" or "fabricates" information that does not directly correspond to the provided input systems and frameworks to monitor the precision and the accuracy of an LLM's output on a continuous basis.

- Interpretability and Explainability: Techniques and measures to make LLMs more transparent and interpretable, enabling stakeholders to understand and trust the decisions made by these models.

- Latency and Inference Time: The computational demands of LLMs can result in increased latency, affecting real-time applications and user experiences. This raises concerns over the applicability of LLMs in areas where timely responses are important.

- Lack of Well-Defined Structures and Frameworks Around Prompt Management: The absence of well-defined structures and frameworks for prompt management is a common challenge in utilizing large language models (LLMs). This crucial aspect of LLM usage often lacks organized tools and established workflows.

What Is an LLMOps Platform?

An LLMOps platform offers a collaborative environment for data scientists and software engineers, enabling them to streamline their workflow. It supports iterative data exploration, tracks experiments, facilitates prompt engineering, manages models and pipelines, and ensures controlled transitioning, deployment, and monitoring of LLMs.

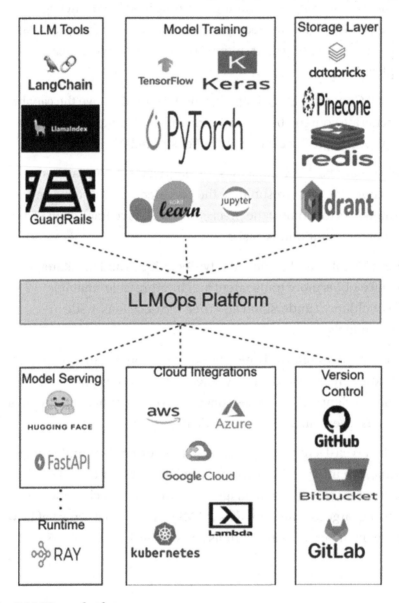

Figure 7-9. *LLMOps platform*

Technology Components LLMOps

Platform/ Framework	Description
Deeplake	Stream large multimodal datasets to achieve near 100% GPU utilization. Query, visualize, and version control data. Access the data without the necessity to recompute the embeddings when performing fine-tuning on the model.
LangFlow	A simple way to experiment and prototype LangChain flows using drag-and-drop components and an intuitive chat interface.
LLMFlows	LLMFlows is a framework for building simple, explicit, and transparent LLM applications such as chatbots, question-answering systems, and agents.
BudgetML	Set up a cost-effective machine learning inference service with a concise code base of fewer than ten lines.
Arize-Phoenix	ML observability for LLMs, vision, language, and tabular models.
ZenML	An open source framework for coordinating, experimenting with, and deploying machine learning solutions suitable for production environments, featuring built-in integrations for LangChain and LlamaIndex.
Dify	This open source framework is designed to empower both developers and nondevelopers to rapidly create practical applications using large language models. It ensures these applications are user-friendly, functional, and capable of continuous improvement.
xTuring	Build and control your personal LLMs with fast and efficient fine-tuning.
Haystack	Creating applications with ease using LLM agents, semantic search, question answering, and additional features.
GPTCache	Establishing a semantic cache for storing responses generated by LLM queries.
EmbedChain	A framework for developing ChatGPT-like bots using your own dataset.

Monitoring Generative AI Models

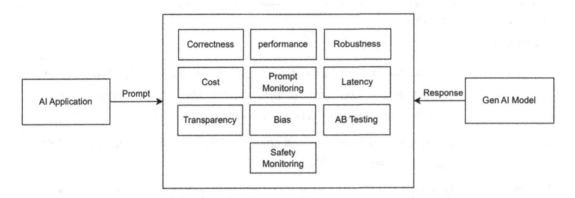

Figure 7-10. *Monitoring generative AI models*

Monitoring generative AI models as shown in Figure 7-10 involves tracking various dimensions to ensure their responsible and effective use. Here's how you can include the aspects of correctness, performance, cost, robustness, prompt monitoring, latency, transparency, bias, A/B testing, and safety monitoring in your monitoring strategy:

1. Correctness:

   - Definition: Correctness refers to the accuracy of the generated content and whether it aligns with the desired outcomes.

   - Monitoring Approach: Use automated validation checks and quality assessments to verify that the generated content is factually accurate and contextually appropriate.

2. Performance:

   - Definition: Performance relates to the quality of generated content in terms of fluency, coherence, and relevance.

   - Monitoring Approach: Continuously measure and analyze performance metrics, such as perplexity, BLEU score, or ROUGE score, to assess the quality of the generated text.

3. Cost:

    •   Definition: Cost monitoring involves tracking the computational resources and infrastructure expenses associated with running the AI model.

    •   Monitoring Approach: Implement cost-tracking tools to monitor resource utilization and optimize costs while maintaining performance.

4. Robustness:

    •   Definition: Robustness assesses the AI model's ability to handle diverse inputs and adapt to different contexts.

    •   Monitoring Approach: Test the model's responses to a wide range of inputs and monitor its behavior under various conditions to ensure it remains reliable.

5. Prompt Monitoring:

    •   Definition: Prompt monitoring involves examining the prompts or inputs provided to the AI model and ensuring they align with ethical guidelines.

    •   Monitoring Approach: Regularly review and audit prompts to prevent misuse or biased inputs.

6. Latency:

    •   Definition: Latency measures the response time of the AI model, ensuring it meets user expectations for timely interactions.

    •   Monitoring Approach: Monitor response times and set latency targets to ensure prompt and efficient interactions.

7. Transparency:

    •   Definition: Transparency involves providing insights into how the AI model operates and makes decisions.

    •   Monitoring Approach: Maintain clear records of model inputs and outputs, and consider implementing transparency tools or techniques like explainable AI to improve model interpretability.

8. Bias:

   - Definition: Bias monitoring focuses on identifying and mitigating biases in the model's outputs, such as gender, race, or cultural biases.

   - Monitoring Approach: Implement bias detection algorithms and conduct regular audits to address and mitigate potential biases in the model's responses.

9. A/B Testing:

   - Definition: A/B testing involves comparing the performance of different model versions or configurations.

   - Monitoring Approach: Conduct A/B tests to assess the impact of changes or updates to the model on user satisfaction, correctness, and other key metrics.

10. Safety Monitoring:

    - Definition: Safety monitoring aims to prevent harmful actions or outputs from the AI model.

    - Monitoring Approach: Implement safety measures, such as content filtering, anomaly detection, and emergency shutdown procedures, to ensure the model operates safely.

    - Consider this example of an "unsafe prompt" related to Indian culture:

    - Unsafe Prompt Example: "Generate a description of Indian cuisine, but focus only on its spiciness and mention that it's too spicy for most people."

    - This prompt is potentially unsafe because it oversimplifies and stereotypes Indian cuisine by reducing it to one aspect (spiciness) and implying that it may be intolerable for many, which is not a fair or accurate representation of Indian food.

    - Monitoring Response: Be vigilant in identifying and rejecting prompts that perpetuate stereotypes, discrimination, or reductionist narratives. Implement bias detection algorithms to flag and address prompts that may lead to inaccurate or biased content. Clearly communicate ethical guidelines that discourage prompts promoting stereotypes or negative generalizations about cultures or cuisines.

- By incorporating these aspects into your monitoring strategy, you can effectively oversee the correctness, performance, cost-efficiency, robustness, promptness, latency, transparency, bias mitigation, A/B testing, and safety of generative AI models. Regularly review and update your monitoring practices to address emerging challenges and ensure responsible AI use.

- This example highlights the importance of monitoring and addressing unsafe prompts that may perpetuate stereotypes or provide an inaccurate representation of cultures, in this case, Indian cuisine.

By incorporating these aspects into your monitoring strategy, you can effectively oversee the correctness, performance, cost-efficiency, robustness, promptness, latency, transparency, bias mitigation, A/B testing, and safety of generative AI models. Regularly review and update your monitoring practices to address emerging challenges and ensure responsible AI use.

Additional Note:

While the section provides a holistic overview of monitoring dimensions for generative AI models, it's worth noting that some readers may find it beneficial to categorize these dimensions based on whether they primarily relate to monitoring the request or the response. This can provide a more granular perspective on the monitoring process and its application within the AI model's workflow.

Readers interested in such a categorization may consider approaching their monitoring strategy by identifying which aspects pertain to incoming requests and which focus on evaluating the AI model's generated responses.

Proprietary Generative AI Models

Proprietary generative AI models are developed by organizations for specific purposes and are typically protected by commercial licensing agreements. They offer advantages in terms of quality, control, and support but may come with usage restrictions and associated costs.

Table 7-1 shows some of the proprietary generative AI models that are available at the moment of writing this book.

Table 7-1. *Generative AI Models Available*

Model	Parameters	Context Length	Fine Tuneable
GPT-3.5	175 billion	4k/16k	Yes
PaLM 2 (Bison)	540 billion	?	No
Cohere	52.4 billion	?	Yes
Claude	175 billion	9k	No
ada, babbage, curie	Up to 7 billion	2k	Yes

Open Source Models with Permissive Licenses

Table 7-2 shows a list of open source models with permissive licenses.

Table 7-2. *Open Source Models*

Language Model	Params	Context Length
T5	11B	2k
UL2	20B	2k
Pythia, Dolly 2.0	12B	2k
MPT-7B	7B	84k
RedPajama-INCITE	7B	2k
Falcon	40B	2k
MPT-30B	30B	8k
LLaMa 2	70B	4k

Playground for Model Selection

A model selection playground as shown in Figure 7-11 is an environment or workspace where data scientists and machine learning practitioners can systematically evaluate and compare different machine learning models and algorithms to choose the most suitable one for a specific task or dataset. Building such a playground involves several steps and considerations, and here is an example of how it could be done.

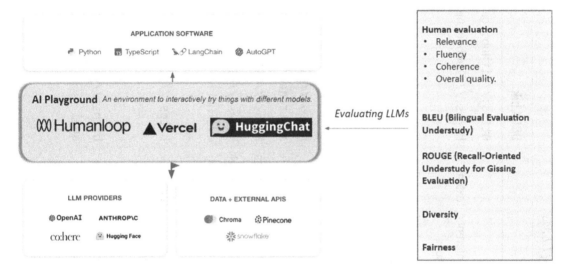

Figure 7-11. *Playground for model selection*

Evaluation Metrics

Evaluation metrics are essential tools in assessing the performance of machine learning models, algorithms, and systems across various tasks. These metrics help quantify how well a model is performing, making it easier to compare different models and make informed decisions. Here are some popular frameworks and libraries for evaluating LLMs:

Table 7-3. Frameworks and Libraries for Evaluating LLMs

Framework Name	Factors Considered for Evaluation	URL Link
Big Bench	Generalization abilities	https://github.com/google/BIG-bench
GLUE Benchmark	Grammar, paraphrasing, text similarity, inference, textual E=entailment, resolving pronoun references	https://gluebenchmark.com/
SuperGLUE Benchmark	Natural language understanding, reasoning, understanding complex sentences beyond training data, coherent and well-formed natural language generation, dialogue with human beings, common sense reasoning, information retrieval, reading comprehension	https://super.gluebenchmark.com/
OpenAI Moderation API	Filter out harmful or unsafe content	https://platform.openai.com/docs/api-reference/moderations
MMLU	Language understanding across various tasks and domains	https://github.com/hendrycks/test
EleutherAI LM Eval	Evaluating and assessing performance across a diverse set of tasks with minimal fine-tuning using a few-shot learning approach	https://github.com/EleutherAI/lm-evaluation-harness
OpenAI Evals	Evaluating the quality and attributes of generated text, including accuracy, diversity, consistency, robustness, transferability, efficiency, and fairness	https://github.com/openai/evals
Adversarial NLI (ANLI)	Robustness, generalization, coherent explanations for inferences, consistency of reasoning across similar examples, efficiency in terms of resource usage	https://github.com/facebookresearch/anli

Name	Description	URL
LIT (Language Interpretability Tool)	Platform to evaluate on user-defined metrics. Insights into their strengths, weaknesses, and potential biases	`https://pair-code.github.io/lit/`
ParlAI	Accuracy, F1 score, perplexity, human evaluation on criteria like relevance, fluency, and coherence, speed and resource utilization, robustness, generalization	`https://github.com/facebookresearch/ParlAI`
CoQA	Understand a text passage and answer a series of interconnected questions that appear in a conversation	`https://stanfordnlp.github.io/coqa/`
LAMBADA	Achieving long-term comprehension by predicting the final word of a given passage	`https://zenodo.org/record/2630551#.ZFUKS-zMLOp`
HellaSwag	Reasoning abilities	`https://rowanzellers.com/hellaswag/`
LogiQA	Logical reasoning abilities	`https://github.com/lgw863/LogiQA-dataset`
MultiNLI	Understanding relationships between sentences across different genres	`https://cims.nyu.edu/~sbowman/multinli/`
SQUAD	Reading comprehension tasks	`https://rajpurkar.github.io/SQuAD-explorer/`

143

Validating LLM Outputs

Validating large language model (LLM) output is a critical step in ensuring the quality, reliability, safety, and ethical use of these powerful language models. Here are some important reasons for validating LLM output:

1. Quality Assurance:

    LLMs are capable of generating a vast amount of text, but not all of it may be of high quality. Validating LLM output helps ensure that the generated content meets desired standards for readability, coherence, and relevance.

2. Ethical Considerations:

    LLMs can sometimes produce content that is biased, offensive, or harmful. Validation is essential to prevent the generation of unethical or inappropriate content, such as hate speech, misinformation, or discriminatory language.

3. Safety:

    To protect users and prevent harm, it's crucial to validate LLM outputs to ensure they do not contain instructions or information that could lead to dangerous actions or self-harm.

4. Bias Mitigation:

    LLMs are known to inherit biases present in their training data. Validating LLM output includes detecting and mitigating biases to ensure fairness and nondiscrimination in the generated content.

5. User Trust:

    Validating outputs helps build and maintain user trust in applications powered by LLMs. Users are more likely to engage with and trust systems that consistently provide high-quality, ethical, and safe content.

6. Compliance with Guidelines:

   Many organizations and platforms have specific guidelines and policies regarding content quality, ethics, and safety. Validation ensures compliance with these guidelines to avoid legal or reputational risks.

7. Continuous Improvement:

   Regularly validating and monitoring LLM output allows for continuous improvement. User feedback and validation results can inform model updates and adjustments to ensure better performance over time.

8. Accountability:

   Keeping records of validation processes and actions taken in response to problematic outputs establishes accountability in case of issues or disputes.

9. Regulatory and Ethical Compliance:

   Compliance with ethical, legal, and regulatory requirements is essential when deploying LLMs in sensitive or regulated domains. Validation helps ensure adherence to these requirements.

10. Customization and Guided Content Generation:

    Validation can be used to guide the LLM's content generation based on specific objectives, allowing organizations to tailor generated content to their needs.

11. Safety Nets:

    Implementing validation mechanisms acts as a safety net to catch and filter out harmful or low-quality content before it is presented to users.

Challenges Faced When Deploying LLMs

1. Computational Resources: Storing and managing the large size of LLMs can be challenging, especially in resource-constrained environments or edge devices. This requires developers to find ways to compress the models or use techniques like model distillation to create smaller, more efficient variants.

2. Model Fine-Tuning: Pre-trained LLMs often need fine-tuning on specific tasks or datasets to achieve optimal performance. This process can be computationally expensive. For example, fine-tuning a 175 Bn parameter DaVinci model would cost $ 180K.

3. Ethical Concerns: LLMs can sometimes generate inappropriate or biased content due to the nature of the data they are trained on. This raises concerns about the ethical implications of deploying such models and the potential harm they might cause.

4. Hallucinations: Hallucinations are a phenomenon in which when users ask questions or provide prompts, the LLM produces responses that are imaginative or creative but not grounded in reality. These responses may appear plausible and coherent but are not based on actual knowledge.

5. Interpretability and Explainability: Understanding the internal workings of LLMs and how making decisions can be difficult due to their complexity. This lack of interpretability poses challenges for developers who need to debug, optimize, and ensure the reliability of these models in real-world applications.

6. Latency and Inference Time: As LLMs have a large number of parameters, they can be slow to generate predictions, particularly on devices with limited computational resources. This can be a challenge when deploying LLMs in real-time applications where low latency is essential.

7. Data Privacy and Access Control: Safeguarding sensitive data used for fine-tuning and inference is crucial. Adhering to data privacy regulations and implementing robust access control mechanisms are paramount to protect user data and maintain trust.

8. Trained Resources for Handling LLMs: Organizations require trained personnel who possess expertise in LLMs, including fine-tuning, ethical considerations, and performance optimization.

9. Model Robustness Across Use Cases: Ensuring that LLMs perform well and provide meaningful responses across diverse applications and domains is a significant challenge as models may excel in some use cases and struggle in others.

10. Legal and Regulatory Compliance: Adhering to legal and regulatory requirements is essential when deploying LLMs, particularly in regulated industries like healthcare and finance. Navigating intellectual property rights, data protection laws, and industry-specific regulations can be intricate.

11. Integration with Existing Systems: Seamlessly integrating LLMs with existing infrastructure and software systems is complex. Compatibility, data flow, and alignment with existing business processes must be carefully considered.

12. Security and Vulnerability Management: Deploying LLMs introduces security risks, including vulnerabilities to adversarial attacks. Developing strategies to identify and mitigate these risks and ensuring secure data transmission are critical.

13. User Feedback Handling: Managing user feedback, particularly in content generation applications, is vital for ongoing model improvement. Establishing mechanisms to process user feedback and incorporate it into model updates is a challenging task.

14. Multilingual and Multimodal Capabilities: If an application necessitates support for multiple languages or multimodal inputs (e.g., text and images), ensuring that the LLM can handle these effectively and provide coherent responses adds complexity to deployment.

15. Long-Term Maintenance: LLM deployment requires continuous maintenance, including monitoring for model drift, adapting to evolving user needs, and addressing emerging challenges.

Implementation

Using the OpenAI API with Python

In today's fast-paced digital landscape, the ability to understand and interact with human language has become a game-changer. OpenAI API emerges as a powerful tool that empowers developers and businesses to seamlessly integrate the prowess of natural language processing into their applications. By tapping into OpenAI's cutting-edge language models, developers can harness the capabilities of AI-driven language understanding, generation, and more.

In this section, we delve into the world of OpenAI API and unveil the steps to effectively leverage its potential using Python. Whether you're crafting intelligent chatbots, generating creative content, or driving insightful language-based interactions, OpenAI API opens doors to endless possibilities. Let's unravel the intricate workings of this API, from setting up your environment to crafting compelling applications that interact intelligently with users. Let's explore the future of human–computer interaction together.

Using the OpenAI API with Python

In this section, we'll walk through the process of using the OpenAI API in Python with a practical example involving the "Alice's Adventures in Wonderland" PDF. We'll explore text generation, analysis, and question answering using the OpenAI API.

Prerequisites

- Python 3.x installed

- Access to OpenAI API and API key

- ChromaDB installation

- Alice's Adventures in Wonderland PDF from `www.gutenberg.org/ebooks/11`

Installation

Firstly, let's install the necessary libraries.

```
!pip install openai langchain pypdf unstructured "unstructured[pdf]"
```

Initializing the Environment and Setting API Key

```
import openai
import langchain
from langchain import OpenAI, VectorDBQA
from langchain.document_loaders import UnstructuredFileLoader
from langchain.text_splitter import CharacterTextSplitter
from langchain.embeddings import OpenAIEmbeddings
from langchain.vectorstores import Chroma
from langchain.chains import RetrievalQA

openai.api_key = "your_openai_api_key_here"
```

Replace "your_openai_api_key_here" with the actual API key you obtained from your OpenAI account.

Test the Environment

Verify that your environment is correctly set up by running a simple API call. For instance, you can try generating text using the "openai.Completion.create()" method.

```
response = openai.Completion.create(
    engine="davinci",
    prompt="Once upon a time in a",
    max_tokens=50
)

print(response.choices[0].text.strip())

land called Wade: Ebner, John Wesley. "Dr. Wade in Hampton," in GRA.

117At a time when no American: Reynolds, Keynote Address.

117To everyone's relief
```

Data Preparation: Loading PDF Data

Load the PDF data.

```
def load_pdf(pdf_path):
    loader = UnstructuredFileLoader(pdf_path)
    pages= loader.load()
    return pages
```

Split the data into chunks:

We are using CharacterTextSplitter to split the PDF content into chunks. Each chunk is then processed separately using the OpenAI API. This approach ensures that the input remains manageable and stays within the token limit while allowing you to analyze or generate text for the entire PDF.

```
pages=load_pdf('/content/drive/MyDrive/Colab Notebooks/Alices_Adventures_in_Wonderland_by_Lewis Carroll.pdf')
text_to_chunks = CharacterTextSplitter(chunk_size=500, chunk_overlap=0)
chunks_of_text = text_to_chunks.split_documents(pages)
```

Remember that the chunk size and overlap can affect the quality and coherence of the results. It's a trade-off between staying within the token limit and maintaining context.

Embeddings and VectorDB Using LangChain and Chroma

LangChain offers a convenient framework for the swift prototyping of local applications based on LLM (large language models). Alongside this, Chroma presents an integrated vector storage and embedding database that seamlessly operates during local developmental stages, empowering these applications.

```
embeddings_function = OpenAIEmbeddings(openai_api_key = openai.api_key)
docsearch = Chroma.from_documents(chunks_of_text,embeddings_function)
chain = RetrievalQA.from_chain_type(llm=OpenAI(), chain_type="stuff", retriever=docsearch.as_retriever())
```

Utilizing OpenAI API

QnA on the PDF:

```
# Input the query at runtime
user_query = input("Enter your query: ")

# Run the QA using the provided query
qa_result = chain.run(user_query)
print("OpenAI Response:", qa_result)
```

```
... Enter your query: [                    ]
```

Query 1: Who is the Hero of this book?

```
# Input the query at runtime
user_query = input("Enter your query: ")

# Run the QA using the provided query
qa_result = chain.run(user_query)
print("OpenAI Response:", qa_result)

Enter your query: Who is the Hero of this book?
OpenAI Response:  The hero of this book is Alice.
```

Query 2: Who is the author of Alice in Wonderland?

```
# Input the query at runtime
user_query = input("Enter your query: ")

# Run the QA using the provided query
qa_result = chain.run(user_query)
print("OpenAI Response:", qa_result)
```

```
Enter your query: Who is the author of Alice in Wonderland?
OpenAI Response:  The author of Alice's Adventures in Wonderland is Lewis Carroll.
```

Query 3: What happens to the size of Alice when she eats or drinks?

```
# Input the query at runtime
user_query = input("Enter your query: ")

# Run the QA using the provided query
qa_result = chain.run(user_query)
print("OpenAI Response:", qa_result)
```

```
Enter your query: What happens to the size of Alice when she eats or drinks?
OpenAI Response:  Alice remains the same size.
```

If you notice the answer to the preceding query is incorrect, OpenAI's response suggests that Alice remains the same size when she eats or drinks. However, in "Alice's Adventures in Wonderland," her size actually changes. This could be due to the context and information available in the specific chunk of text that was analyzed. Keep in mind that the accuracy of the response depends on the content and context of the text being processed by the OpenAI model.

Note rewriting the query with more context gives us a better result.

```
[71] # Input the query at runtime
     user_query = input("Enter your query: ")

     # Run the QA using the provided query
     qa_result = chain.run(user_query)
     print("OpenAI Response:", qa_result)

     Enter your query: Did Alice grow larger in the book? If yes, why?
     OpenAI Response:  Yes, Alice grows larger in the book. She grew larger because she ate a piece of mushroom which caused her to grow in size.
```

Query 4: Analyze the interactions between Alice and the Queen of Hearts in the PDF.

```
# Input the query at runtime
user_query = input("Enter your query: ")

# Run the QA using the provided query
qa_result = chain.run(user_query)
print("OpenAI Response:", qa_result)
```

```
Enter your query: Analyze the interactions between Alice and the Queen of Hearts in the PDF
OpenAI Response:
Alice and the Queen of Hearts have a tense exchange in the PDF. Alice is bold and unafraid to voice her opinion, even when interrupting the King.
```

In conclusion, this guide demonstrates the integration of the OpenAI API, LangChain, and ChromeDb to extract insights from the "Alice in Wonderland" PDF and perform targeted queries. This combination of contemporary technology with classic literature offers a unique and innovative approach, showcasing the power of modern tools in the analysis of timeless tales.

Leveraging Azure OpenAI Service

The Azure OpenAI Service offers convenient REST API access to a selection of robust language models, including the highly advanced GPT-4, GPT-35-Turbo, and the Embeddings model series. Furthermore, it is worth noting that the GPT-4 and gpt-35-turbo model series are now available for general use. These models can be seamlessly tailored to suit your specific needs, encompassing tasks such as content creation, summarization, semantic search, and natural language-to-code translation. You can engage with the service via REST APIs, the Python SDK, or through our web-based interface available in the Azure OpenAI Studio.

Moreover, one of the key advantages of leveraging the Azure OpenAI Service is the ability to seamlessly swap language models based on your requirements. This swappability becomes even more potent when integrated with orchestrators like LangChain. With this setup, you can easily switch between different language models to suit specific tasks or scenarios. Whether you need a model for content generation, language translation, or any other natural language processing task, the combination of swappable LLMs and orchestrators provides the adaptability your enterprise needs.

Implementing the Azure OpenAI Service into your enterprise's workflow can unlock new possibilities for natural language understanding, generation, and interaction. It's a powerful tool for enhancing customer experiences, automating processes, and gaining insights from textual data. Please find the following URL for a detailed guide on how to implement Azure AI for your enterprise on the Microsoft Azure website.

URL: `https://azure.microsoft.com/en-us/solutions/ai`

Conclusion

In the ever-evolving landscape of enterprise technology, large language models (LLMs) have emerged as formidable allies, offering a profound transformation in how businesses operate, interact, and innovate. As we conclude this chapter, we find ourselves at the intersection of opportunity and innovation, where the power of LLMs converges with the ambitions of forward-thinking enterprises.

Throughout this chapter, we have explored three compelling approaches for harnessing the capabilities of LLMs within enterprise settings:

Private Generalized LLM API: We delved into the concept of a private generalized LLM API, highlighting the value it brings through data privacy, customization, and control. We witnessed how it empowers enterprises to sculpt tailored solutions, amplify customer engagement, and navigate the intricate terrain of natural language interactions. By incorporating this approach, enterprises stand poised to create transformative experiences while safeguarding sensitive data.

Context Injection Architecture: We ventured into the realm of context injection architecture, an ingenious strategy to augment LLMs with domain-specific knowledge and context. As we explored its potential, we unveiled how it enhances customer support, elevates content curation, and sharpens decision-making processes. Enterprises that embrace this approach can fortify their offerings, providing clients and users with enriched, context-aware interactions.

Fine-Tuning LLMs for Enterprise Use Cases: The concept of fine-tuning LLMs opened doors to precision and adaptability. We observed how this practice elevates LLMs by optimizing their accuracy, imbuing them with domain-specific language, and enhancing their task-specific performance. In scenarios spanning sentiment analysis, legal document review, and code generation, enterprises can leverage fine-tuned LLMs to achieve unparalleled outcomes tailored to their unique needs.

As we reflect on these approaches, we are reminded that the journey with LLMs is not a destination but an ongoing exploration. In a world where technology evolves ceaselessly, enterprises that embrace LLMs and adapt to their potential are better equipped to tackle the challenges and seize the opportunities that lie ahead.

The marriage of LLMs and enterprise solutions is not merely a glimpse into the future; it is a bold step toward shaping it. The possibilities are boundless, and the path forward promises innovations yet unimagined. We invite enterprises to embark on this transformative journey, armed with the knowledge and strategies to harness the full potential of LLM technology.

As we move into an era where language models are more than tools—they are partners in innovation—enterprises that embrace LLMs will not only navigate the future but also lead the way, ushering in an era of enriched customer experiences, streamlined operations, and uncharted possibilities. The journey has begun, and the future is in our hands.

CHAPTER 8

Diffusion Model and Generative AI for Images

The two prominent generative models, namely, generative adversarial networks (GANs) and variational autoencoders (VAEs), have gained substantial recognition. We will see a brief explanation of both in this chapter followed by a detailed diffusion model. GANs have exhibited versatility across various applications, yet their training complexity and limited output diversity, caused by challenges like mode collapse and gradient vanishing, have been evident. On the other hand, VAEs, while having a strong theoretical foundation, encounter difficulties in devising effective loss functions, resulting in suboptimal outputs.

Another category of techniques, inspired by probabilistic likelihood estimation and drawing parallels from physical phenomena, has emerged—these are known as diffusion models. The core concept of diffusion models is rooted in principles similar to the movement of gas molecules in thermodynamics, where molecules disperse from regions of high density to low density, representing an increase in entropy or heat dissipation. In the realm of information theory, this relates to the progressive introduction of noise leading to information loss.

At the heart of diffusion modeling lies the intriguing notion that if we can construct a learning model capable of capturing the gradual degradation of information due to noise, it should theoretically be feasible to reverse this process, thereby reclaiming the original information from the noise. This concept bears a resemblance to VAEs, wherein an objective function is optimized by projecting data into a latent space and subsequently recovering it to its initial state. However, the distinction lies in the fact that diffusion models don't strive to learn the data distribution directly. Instead, they focus on modeling a series of noise distributions within a Markov chain framework, effectively "decoding" data by iteratively removing noise in a hierarchical manner.

© Akshay Kulkarni, Adarsha Shivananda, Anoosh Kulkarni, Dilip Gudivada 2023
A. Kulkarni et al., *Applied Generative AI for Beginners*, https://doi.org/10.1007/978-1-4842-9994-4_8

Before jumping into diffusion models, let us see a brief explanation of VAEs and GANs.

Variational Autoencoders (VAEs)

Variational autoencoders (VAEs) are a type of generative model that combines ideas from autoencoders and probabilistic modeling. VAEs are designed to learn a latent representation of data that captures meaningful features while also generating new data samples that resemble the original dataset. They are particularly useful for tasks like data compression, denoising, and generative modeling:

1. Encoder: The encoder part of the VAE takes input data and maps it to a latent space. Unlike traditional autoencoders, the encoder of a VAE doesn't produce a fixed encoding but instead outputs a probability distribution over the latent variables. This allows VAEs to capture uncertainty in the encoding process.

2. Latent Space: The latent space is a lower-dimensional representation of the input data. Each point in this space corresponds to a potential data sample. VAEs assume that the data in the latent space follows a specific probabilistic distribution, often a Gaussian distribution.

3. Reparameterization Trick: To enable backpropagation for training, VAEs use a reparameterization trick. Instead of directly sampling from the latent distribution, a sample is generated by adding random noise to the mean and standard deviation parameters of the distribution. This makes it possible to compute gradients for training.

4. Decoder: The decoder takes a sample from the latent space and maps it back to the original data space. Like the encoder, the decoder also outputs a probability distribution over the data, allowing the model to capture uncertainty in the generation process.

5. Loss Function: VAEs are trained to maximize a lower bound on the data likelihood. This lower bound consists of two terms: a reconstruction loss that measures how well the generated data matches the original data and a regularization term that encourages the latent distribution to resemble the assumed prior distribution. The regularization term helps in ensuring that the latent space remains structured and continuous.

6. Generation and Interpolation: Once trained, a VAE can generate new data samples by sampling from the latent space and passing the samples through the decoder. Additionally, because the latent space has a smooth structure, interpolations between points in this space result in meaningful interpolations in the data space.

VAEs have demonstrated their effectiveness in various applications, including image generation, data compression, and domain adaptation. They provide a principled way to learn meaningful latent representations of data while generating diverse and realistic new samples. However, VAEs might produce slightly blurry outputs compared to other generative models like GANs due to the inherent trade-off between reconstruction accuracy and sample diversity in their objective function.

Generative Adversarial Networks (GANs)

Generative adversarial networks (GANs) are a class of machine learning models designed to generate new data that is similar to a given dataset. GANs consist of two main components: a generator and a discriminator. The generator creates synthetic data samples, while the discriminator evaluates these samples and tries to distinguish between real and generated data. The two components are trained together in a competitive process, leading to the refinement of both the generator's ability to create realistic data and the discriminator's ability to differentiate between real and fake data:

1. Generator (G): The generator takes random noise as input and transforms it into data that should resemble the target dataset. Initially, its output might not resemble the real data much.

2. Discriminator (D): The discriminator acts as a binary classifier. It takes both real data from the target dataset and generated data from the generator as input and tries to determine whether the input is real (from the dataset) or fake (generated by the generator).

3. Training Process: The training of GANs involves an adversarial process. The generator and discriminator are trained iteratively. During each iteration:

   - The generator generates fake data from random noise.

   - The discriminator is given real data and the generated fake data, and it learns to distinguish between them.

   - The generator's parameters are adjusted to produce better fake data that the discriminator struggles to differentiate from real data.

4. Objective: The goal of the generator is to improve its ability to produce data that is so convincing that the discriminator cannot distinguish it from real data. The goal of the discriminator is to become better at correctly classifying real and fake data.

5. Equilibrium: As training progresses, the generator and discriminator reach a point of equilibrium where the generator generates data that is increasingly difficult for the discriminator to distinguish from real data. This results in the generation of high-quality synthetic data.

GANs have been used for various applications, including image synthesis, style transfer, super-resolution, data augmentation, and more. They have shown the capability to create highly realistic data samples and have been responsible for impressive advancements in generative modeling and computer vision. However, GANs can be challenging to train due to issues like mode collapse (when the generator focuses on a limited subset of the target data) and training instability.

Diffusion Models

Diffusion models are a relatively novel class of generative models that draw inspiration from physical processes like the diffusion of particles and concepts from information theory. They aim to generate data by iteratively transforming noise into structured information, essentially reversing the process of noise introduction.

In a nutshell, diffusion models work as follows:

1. Noise Schedule: A sequence of noise levels is defined, gradually increasing from minimal noise to more significant noise. Each noise level represents a trade-off between clarity and noise in the data.

2. Markov Chain: Diffusion models utilize a Markov chain, which consists of multiple steps corresponding to the different noise levels in the schedule. At each step, the model processes the data by adding noise and gradually distorting it.

3. Conditional Modeling: The model creates a conditional distribution that estimates what the data looks like at each noise level, given the data at the previous level. This effectively captures the degradation of the data due to noise.

4. Reverse Process: After the data has been processed through the Markov chain with increasing noise levels, a reverse process is applied. This process aims to recover the original data by iteratively removing the noise, moving back through the noise schedule.

5. Training Objective: Diffusion models are trained by optimizing the parameters to minimize the difference between the estimated data distributions at each noise level and the actual data observed at those levels. This is typically achieved by maximizing the likelihood of observing the data given the modeled diffusion process.

The concept behind diffusion models is to represent the gradual loss of information due to noise and then use this knowledge to recover the original information by undoing the noise introduction. Unlike traditional generative models that directly model the data distribution, diffusion models focus on modeling the process of noise addition and removal.

Diffusion models have shown promise in generating high-quality data samples with diverse characteristics. They hold the potential to capture complex data distributions and handle scenarios where the quality of the data degrades over time, which can be particularly useful for applications in image generation, data denoising, and more.

However, as of my last update in September 2021, diffusion models might not be as widely studied or implemented as other generative models like GANs or VAEs.

Types of Diffusion Models

There are many different types of diffusion models, but some of the most common include as follows:

- Denoising Diffusion Probabilistic Models (DDPMs): DDPMs are a type of diffusion model that starts with a noisy image and gradually removes the noise to reveal the underlying image. DDPMs are trained using a technique called maximum likelihood estimation, which means that they are trained to minimize the distance between the generated images and the real images in the training dataset.

    Figure 8-1 illustrates denoising diffusion probabilistic models.

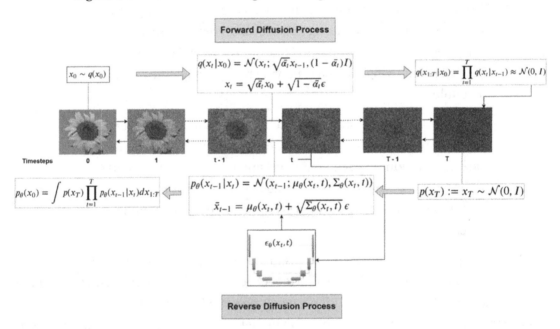

Figure 8-1. *Denoising diffusion probabilistic models (DDPMs)*

Reference: `https://learnopencv.com/wp-content/`
`uploads/2023/02/denoising-diffusion-probabilistic-`
`models-forward_and_backward_equations-1536x846.png`

- Score-Based Diffusion Models (SBMs): SBMs are a type of diffusion model that uses a score function to generate images. The score function is a function that measures how likely an image is to be real. SBMs are trained using a technique called adversarial training, which means that they are trained to generate images that are indistinguishable from real images.

Figure 8-2 illustrates score-based diffusion models.

Score Based
Generative Models

$$s(\mathbf{x}) \triangleq \nabla_{\mathbf{x}} \log p_{\text{data}}(\mathbf{x})$$

Figure 8-2. *Score-based diffusion models*

- Stochastic Differential Equation (SDE)-Based Diffusion Models: SDE-based diffusion models are a type of diffusion model that uses a stochastic differential equation (SDE) to generate images. SDEs are equations that describe the evolution of a random process over time. SDE-based diffusion models are trained using a technique called generative adversarial training, which means that they are trained to generate images that are indistinguishable from real images.

Figure 8-3 illustrates stochastic differential equation (SDE)-based diffusion models.

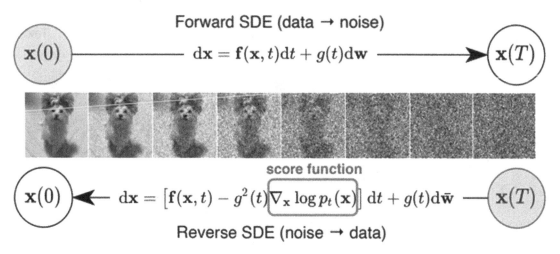

Figure 8-3. *Stochastic differential equation (SDE)-based diffusion models*

Diffusion models have been used successfully for a variety of tasks, including the following:

- Image Generation: Diffusion models can be used to generate realistic images from text descriptions.

- Text-to-Image Synthesis: Diffusion models can be used to synthesize images from text descriptions.

- Style Transfer: Diffusion models can be used to transfer the style of one image to another image.

- Super-resolution: Diffusion models can be used to super-resolve low-resolution images.

Architecture

Diffusion models are a powerful tool for generating realistic and creative content. They are still under development, but they have the potential to revolutionize the way we create and interact with images.

The architecture of diffusion models is relatively simple. They consist of two main components.

Figure 8-4 illustrates latent representation model in diffusion models.

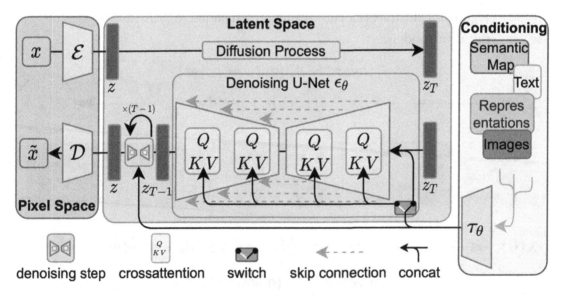

Figure 8-4. *Latent representation model in diffusion models*

- Latent Representation Model: The latent representation model is typically a neural network that takes an image as input and outputs a latent representation of the image. The latent representation is a vector of numbers that captures the essential features of the image. The latent representation model is trained on a dataset of real images. The goal of the latent representation model is to learn a mapping from images to latent representations such that images that are similar to each other have similar latent representations.

The latent representation model can be implemented using any type of neural network, but convolutional neural networks (CNNs) are often used. CNNs are well-suited for image processing tasks because they can learn to extract features from images at different scales.

The latent representation model is trained using a technique called maximum likelihood estimation. Maximum likelihood estimation is a statistical technique that finds the parameters of a model that maximize the likelihood of the observed data. In the case of the latent representation model, the observed data is the dataset of real images. The goal of maximum likelihood estimation is to find the parameters of the latent representation model that make the model most likely to have generated the real images in the dataset.

Figure 8-5 illustrates the diffusion process in diffusion models.

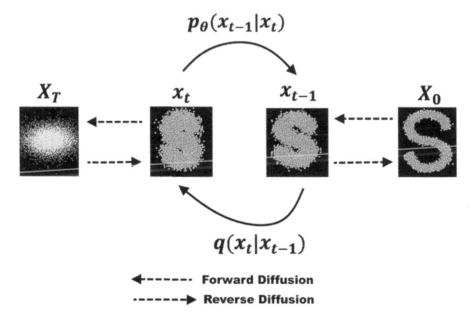

Figure 8-5. *Diffusion process in diffusion models*

- Diffusion Process: The diffusion process is a Markov chain that takes a latent representation as input and gradually modifies it to generate a new image. The diffusion process is a probabilistic process, which means that it can only move from one state to the next in a certain way. The diffusion process is trained to generate images that are indistinguishable from real images.

The diffusion process works by first adding noise to the latent representation. The amount of noise that is added is determined by a parameter called the diffusion rate. The diffusion rate is gradually increased as the diffusion process progresses. This means that the generated images become more and more different from the original image as the diffusion process progresses.

The diffusion process can be implemented using any type of Markov chain, but a common approach is to use a Gaussian diffusion process. A Gaussian diffusion process is a Markov chain that adds Gaussian noise to the latent representation at each step.

The diffusion process is trained using a technique called adversarial training. Adversarial training is a technique for training generative models that pits two models against each other. In the case of diffusion models, the two models are the diffusion process and a discriminator. The discriminator is a neural network that is trained to distinguish between real images and generated images.

The goal of adversarial training is to train the diffusion process to generate images that are so realistic that the discriminator cannot distinguish them from real images. This is done by iteratively updating the parameters of the diffusion process and the discriminator until the discriminator is unable to distinguish between real images and generated images with high confidence.

- Decoding Process: The decoding process is typically a neural network that takes a latent representation as input and outputs an image. The decoding process is trained to reconstruct the original image from the latent representation.

The decoding process can be implemented using any type of neural network, but CNNs are often used. CNNs are well-suited for image reconstruction tasks because they can learn to invert the operations that were performed by the latent representation model.

The decoding process is trained using a technique called mean squared error (MSE) loss. MSE loss is a loss function that measures the difference between the reconstructed image and the original image. The goal of MSE loss is to minimize the difference between the reconstructed image and the original image.

In recent years, the field of artificial intelligence (AI) has witnessed significant progress, introducing various innovations. One notable addition to the AI landscape is the emergence of AI image generators. These sophisticated tools possess the capability to transform textual input into vivid images or artistic depictions. Among the plethora of options available for text-to-image AI solutions, several have garnered particular attention, the ones that stand out are DALL-E 2, stable diffusion, and Midjourney.

The Technology Behind DALL-E 2

Have you ever been curious about how AI is capable of turning words into images? Imagine describing something in text and then witnessing AI craft an image of that description. Generating high-quality images solely from textual descriptions has posed a significant challenge for AI researchers. This is precisely where DALL-E and its advanced version, DALL-E 2, come into play. In this article, we're delving into the intricacies of DALL-E 2.

Developed by OpenAI, DALL-E 2 is an advanced AI model with the remarkable ability to produce remarkably realistic images based on textual descriptions. But how does DALL-E 2 achieve this feat, and what sets it apart? Throughout this post, we're delving into the fundamental concepts and techniques underpinning DALL-E 2. We'll explore concepts like contrastive language-image pre-training (CLIP), diffusion models, and postprocessing. Moreover, we'll touch on the computational resources necessary for training a model like DALL-E 2, along with the deep learning frameworks and libraries that facilitate its implementation. By the time you've finished reading, you'll have a solid grasp of how DALL-E 2 operates and what makes it a groundbreaking advancement in the realm of generative AI.

DALL-E 2 represents an evolved version of the original DALL-E, operating within the domain of large language models. This generative model utilizes the power of diffusion models to transform textual descriptions into tangible images. It takes advantage of an encoder-decoder architecture, with a distinctive workflow centered around contrastive language-image pre-training (CLIP) embeddings:

1. Input Text Processing:

   At the start, DALL-E 2 takes in textual descriptions provided by users, describing the image they envision.

2. Encoding Using CLIP:

   The input text undergoes encoding using the CLIP neural network. CLIP is adept at transforming both text and image inputs into high-dimensional embeddings, capturing their semantic essence. This results in a vector representation termed CLIP text embeddings, encapsulating the textual description's meaning.

3. Conversion to CLIP Image Embeddings via Prior:

   The CLIP text embeddings are then directed through a "Prior," which can be either an autoregressive or a diffusion model. This is a critical step where the transition from text to image takes place.

   The Prior, operating as a generative model, harnesses a probability distribution to craft lifelike images. Specifically, the diffusion model is favored due to its superior performance in generating high-quality images.

4. Final Image Generation:

   Once the Prior, particularly the diffusion model, yields CLIP image embeddings, these embeddings are conveyed to the diffusion decoder.

   The diffusion decoder's role is to translate these embeddings into the ultimate image, bringing to fruition the visual representation described in the input text.

   Importantly, there was experimentation during DALL-E 2's development. While a direct approach of using CLIP text embeddings in the decoder (step #4) was tried, integrating a prior (step #3) turned out to be more effective in enhancing image generation quality.

   DALL-E 2's distinctive process allows it to turn textual descriptions into intricate and meaningful images, showcasing the remarkable progress at the crossroads of language and image generation.

Figure 8-6 illustrates *DALL-E 2*.

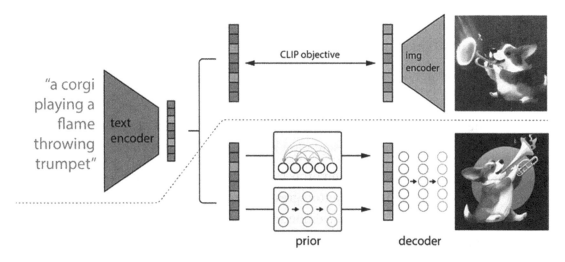

Figure 8-6. *DALL-E 2*

The visual diagram provided illustrates the following concepts:

Top Part: CLIP Training Process

- The upper portion of the image depicts the CLIP training process. CLIP refers to contrastive language-image pre-training.

- This stage involves training a model that learns a shared representation space for both textual and image data.

- The result is a joint representation space where text and images are embedded, allowing them to be compared and related in a meaningful way.

- This shared representation space forms the foundation for understanding the connection between textual descriptions and corresponding images.

Bottom Part: Text-to-Image Generation Process

- The lower part of the image represents the process of transforming text descriptions into images using DALL-E 2.

- The text input, which describes the desired image, is fed into DALL-E 2.

- The input text is encoded using the CLIP encoder, generating a high-dimensional vector representation known as CLIP text embeddings.

- These embeddings are then processed through a Prior, which is a generative model (either autoregressive or diffusion model). The Prior generates CLIP image embeddings, capturing the visual content corresponding to the textual description.

- Finally, these CLIP image embeddings are decoded by the diffusion decoder to produce the final image that aligns with the provided text description.

Visually connects the training of CLIP for joint text-image representation (top part) with the subsequent process of generating images from text using DALL-E 2 (bottom part). It highlights the relationship between the learned embeddings and the conversion of these embeddings into concrete images, showcasing the interplay between textual and visual information within the context of DALL-E 2's operations.

The Technology Behind Stable Diffusion

Stable diffusion is grounded in a sophisticated technology known as latent diffusion model (LDM). This technology constitutes the core of stable diffusion's approach to text-to-image synthesis. Let's explore the technology behind stable diffusion:

Latent Diffusion Model (LDM)

LDM forms the backbone of stable diffusion's methodology. It leverages the principles of diffusion models and their application within the latent space of pre-trained autoencoders. The technology involves several key components and concepts:

1. Diffusion Models in Latent Space:

   - Diffusion models, which gradually transform input data by adding noise and then attempt to reconstruct the original data, are adapted to operate within the latent space.

   - Instead of applying diffusion directly to the input data (like images), diffusion is applied in the latent space of autoencoders. This introduces noise to the latent representations of data.

2. Autoencoders and Latent Representations:

   - Autoencoders are neural networks designed to encode input data into a compressed latent representation and decode it back to the original data.

   - In the context of LDM, the latent space of powerful pre-trained autoencoders is utilized. This latent space captures meaningful features of the input data.

3. Training and Optimization:

   - The LDM is trained to learn the transformation of latent representations under the diffusion process.

   - The training involves optimizing the model's parameters to ensure that the diffusion process effectively captures the noise introduction and subsequent denoising in the latent space.

4. Cross-Attention Layer:

   - An essential augmentation in LDM architecture is the incorporation of a cross-attention layer.

- This layer enhances the model's ability to handle various conditional inputs, such as text descriptions and bounding boxes.

- It plays a pivotal role in facilitating high-resolution image synthesis through convolution-based methods.

Benefits and Significance

- Computational Efficiency: LDMs offer the advantage of training diffusion models on limited computational resources by utilizing the latent space of pre-trained autoencoders.

- Complexity and Fidelity: By training diffusion within the latent space, LDMs strike a balance between simplifying the representation and preserving intricate details, resulting in enhanced visual fidelity.

- Conditioned Synthesis: The integration of a cross-attention layer empowers LDMs to generate images conditioned on diverse inputs like text, contributing to their versatility.

Stable diffusion harnesses the potential of latent diffusion models to create an innovative framework that combines the power of diffusion models, latent representations, and conditioned synthesis. This technology exemplifies the continual evolution of AI-driven image synthesis methods, offering an efficient and effective approach to creating compelling visuals from textual descriptions.

The Technology Behind Midjourney

Midjourney employs a sophisticated technology to facilitate its text-to-image generation capabilities. Let's delve into the underlying technology behind Midjourney:

Generative Adversarial Networks (GANs)

- GANs consist of two components: a generator and a discriminator. The generator crafts images based on random noise, while the discriminator attempts to differentiate between real images and those generated by the generator.

- This adversarial process compels the generator to continually improve its image generation to fool the discriminator.

Text-to-Image Synthesis with GANs

- Midjourney leverages the GAN architecture to synthesize images from textual descriptions.

- The generator is conditioned on text inputs, ensuring that the generated images align with the provided descriptions.

- The text input is usually encoded into a latent representation that guides the image generation process.

Conditional GANs

- Midjourney employs a variant of GANs known as conditional GANs (cGANs).

- In cGANs, both the generator and discriminator are conditioned on additional information (in this case, the text description).

- The conditioning enhances the generator's ability to create images that correspond to specific text prompts.

Training Process

- Midjourney's training process involves iteratively updating the generator and discriminator components.

- The generator aims to create images that the discriminator cannot distinguish from real ones, while the discriminator aims to improve its discrimination ability.

Loss Functions and Optimization

- Loss functions play a crucial role in guiding the training process.

- The generator and discriminator are optimized using specific loss functions that capture the quality of generated images and the discriminator's discrimination accuracy.

Attention Mechanisms

- Midjourney's technology might incorporate attention mechanisms to enhance the generator's focus on relevant parts of the image.

- Attention mechanisms enable the model to selectively emphasize certain regions based on the input text, contributing to more contextually relevant image generation.

Data Augmentation and Preprocessing

- Midjourney might employ data augmentation techniques to expand the training dataset and improve generalization.

- Preprocessing of textual descriptions might involve techniques like tokenization and embedding to convert text into a format suitable for the model.

Benefits and Applications

- Midjourney's technology enables the creation of realistic images based on textual descriptions, making it valuable for various applications like design, content creation, and visualization.

In essence, Midjourney's technology capitalizes on the power of GANs, especially conditional GANs, to transform textual inputs into compelling and contextually relevant images. This approach showcases the synergy between language and image synthesis, opening up avenues for innovative applications in the realm of generative AI.

Comparison Between DALL-E 2, Stable Diffusion, and Midjourney

1. DALL-E 2

   - Training Data: Trained on millions of stock images, resulting in a sophisticated output suitable for enterprise applications.

- Image Quality: Known for producing high-quality images, particularly excelling when generating complex scenes with more than two characters.

- Use Case: Well-suited for enterprise-level usage due to its refined output quality.

- Artistic Style: While capable of generating various styles, DALL-E 2 emphasizes accuracy and realism.

- Access: Availability and access details aren't specified.

2. Midjourney:

- Artistic Style: Renowned for its artistic style, producing images that resemble paintings rather than photographs.

- Operation: Utilizes a Discord bot for sending and receiving calls to AI servers, making interactions happen within the Discord platform.

- Image Output: Primarily generates artistic and creative visuals, aligning with its emphasis on artistic expression.

- Use Case: Ideal for artistic and creative endeavors but might not be optimized for realistic photo-like images.

- Access: Usage details and accessibility aren't explicitly mentioned.

3. Stable Diffusion:

- Open Source: Accessible to a wide audience as an open source model.

- Artistic Understanding: Demonstrates a good grasp of contemporary artistic illustration, producing intricate and detailed artwork.

- Image Creation: Particularly excels at generating detailed and creative illustrations, less suitable for creating simple images like logos.

- Complex Prompts: Requires clear interpretation of complex prompts for optimal results.

- Use Case: Well-suited for creative illustrations and detailed artwork.

- Access: Accessible to a broad user base due to its open source nature.

In summary:

- DALL-E 2 stands out for its enterprise-grade output quality and ability to generate complex scenes with accuracy.

- Midjourney is notable for its artistic and creative style, often producing images resembling paintings.

- Stable diffusion is versatile, offering detailed artistic illustration and creative output, especially for complex prompts.

The choice among these tools depends on the specific use case, desired image style, and the level of detail required. Each tool has its unique strengths, making them suitable for various creative and practical applications.

Applications

Image generator AI tools have a wide range of applications across various industries and domains. Here are some notable applications:

1. Content Creation and Design:

   - These tools can be used to generate visual content for websites, social media, advertisements, and marketing campaigns.

   - Designers can quickly create images to accompany articles, blog posts, and other written content.

2. Concept Visualization:

   - Architects and designers can use these tools to bring concepts to life by generating images based on textual descriptions of buildings, interiors, and landscapes.

3. Art and Entertainment:

   – Artists can use these tools to turn their imaginative ideas expressed in text into actual visual artworks.

   – Video game developers can create scenes, characters, and assets based on written game descriptions.

4. Fashion and Product Design:

   – Designers can generate visual representations of clothing, accessories, and other products before producing physical prototypes.

5. Storytelling and Literature:

   – Authors can use these tools to illustrate scenes from their stories or create visual prompts for inspiration.

   – Comics and graphic novel creators can translate scripts into visuals.

6. Educational Materials:

   – Teachers and educators can use these tools to generate images for educational materials and presentations.

   – Visual aids can enhance learning by providing concrete examples for abstract concepts.

7. Ecommerce and Catalogs:

   – Ecommerce platforms can automatically generate product images from textual descriptions, aiding in catalog creation.

8. Prototype Visualization:

   – Engineers and product developers can quickly visualize prototypes based on written specifications, aiding in the design process.

9. Medical Imaging and Visualization:

   – Medical professionals can generate visual representations of medical conditions, aiding in patient education and communication.

10. Creative Advertising:

   – Advertisers can create unique and engaging visuals for campaigns based on written creative briefs.

11. Interior Design:

   – Interior designers can visualize and experiment with different design ideas based on text descriptions before implementing them.

12. Cinematography and Storyboarding:

   – Filmmakers and animators can use these tools to create storyboards and previsualize scenes.

13. Research Visualization:

   • Researchers can visualize complex data and research findings, making them more accessible to a broader audience.

14. Fashion Forecasting:

   • Fashion industry professionals can generate images of potential fashion trends based on text descriptions and predictions.

15. Automated Art Generation:

   • Artists can use these tools to generate new and unique artworks, exploring novel styles and compositions.

These applications highlight the versatility of diffusion models and text-to-image generator AI tools, demonstrating their potential to transform textual descriptions into valuable visual assets across diverse fields.

Conclusion

The realm of image creation tools has witnessed remarkable evolution, with diffusion models and text-to-image generator AI tools standing at the forefront of innovation. Diffusion models, inspired by physical processes, offer a novel approach to generating images by adding noise and subsequently reconstructing the original data. These models, whether employed independently or within the latent space of autoencoders, strike a delicate balance between complexity reduction and detail preservation. The incorporation of cross-attention layers further empowers diffusion models, enabling them to cater to diverse conditional inputs and yielding high-resolution, contextually relevant outputs.

Text-to-image generator AI tools, such as DALL-E 2, stable diffusion, and Midjourney, embody diverse strategies for transforming textual descriptions into vivid visual representations. Each tool has distinct strengths, from DALL-E 2's enterprise-grade output quality to stable diffusion's accessibility and Midjourney's emphasis on artistic expression. These tools not only bridge the gap between language and visual content but also pave the way for novel applications across industries. From content creation and design to architecture, entertainment, education, and more, the applications of these tools are far-reaching and diverse.

As the field continues to advance, diffusion models and text-to-image generator AI tools are poised to redefine creativity, design, and communication. Their ability to harness the power of language and imagery has the potential to transform industries, enhance user experiences, and inspire new forms of expression. With ever-improving technologies and expanding use cases, the future promises exciting possibilities at the intersection of AI, image generation, and human creativity.

## CHAPTER 9

ChatGPT Use Cases

In the era of GenAI, ChatGPT stands as a remarkable and versatile tool with myriad applications across diverse domains. From transforming the landscape of business and customer service to revolutionizing content creation, marketing strategies, and language and communication tasks, ChatGPT's capabilities transcend traditional boundaries. It plays a pivotal role in software development, healthcare, market research, creative writing, education, legal compliance, HR functions, and data analysis, demonstrating its immense potential in shaping the way we approach complex challenges and decision-making across various sectors. This exploration delves into the multifaceted use cases of ChatGPT across different domains, shedding light on its remarkable adaptability and impact.

Business and Customer Service

1. Customer Support:

    ChatGPT can revolutionize customer support by providing instant, round-the-clock assistance. It handles a wide range of customer queries, from simple FAQ inquiries to complex troubleshooting issues. Through its natural language understanding and generation capabilities, ChatGPT engages in humanlike conversations, ensuring customers receive timely and accurate responses.

    Example: A customer contacts an ecommerce website with a question about a product's specifications. ChatGPT understands the query, retrieves the relevant information from its knowledge base, and delivers a detailed response to the customer's satisfaction.

© Akshay Kulkarni, Adarsha Shivananda, Anoosh Kulkarni, Dilip Gudivada 2023
A. Kulkarni et al., *Applied Generative AI for Beginners*, https://doi.org/10.1007/978-1-4842-9994-4_9

2. Sales and Product Information:

    ChatGPT becomes a virtual sales assistant, offering customers information about products and services. It assists in decision-making by providing detailed descriptions, specifications, and pricing, and even suggesting related products based on customer preferences.

    Example: A potential buyer is exploring laptops on an electronics website. ChatGPT engages in a conversation, asking about the buyer's requirements and preferences. It then recommends laptops that match the buyer's needs and provides a comparison of their features.

3. Feedback Analysis and Improvement:

    Businesses can use ChatGPT to analyze customer feedback and sentiment. By processing reviews, comments, and surveys, ChatGPT provides insights into customer perceptions, helping companies identify areas for improvement and fine-tuning their products and services.

    Example: A restaurant chain uses ChatGPT to analyze customer reviews. It detects recurring mentions of slow service and subpar presentation. The restaurant management takes action to address these issues, leading to improved customer satisfaction.

4. Personalized Recommendations:

    ChatGPT can offer personalized recommendations to customers based on their preferences and behavior. By analyzing past interactions and purchase history, it suggests products or services that align with the customer's interests.

    Example: A user is browsing an online clothing store. ChatGPT suggests outfits and accessories that match the user's style based on their previous purchases and browsing history.

5. Order Tracking and Status Updates:

   Customers often seek information about their orders' status and tracking details. ChatGPT handles these inquiries by providing real-time updates on shipping, delivery times, and any delays.

   Example: A customer inquires about the status of their online order. ChatGPT retrieves the latest tracking information and informs the customer that the package is out for delivery, along with an estimated arrival time.

6. Handling Returns and Refunds:

   ChatGPT assists customers in initiating returns or requesting refunds by guiding them through the process. It explains return policies, provides instructions for packaging items, and helps generate return labels.

   Example: A customer wants to return a defective product purchased online. ChatGPT guides the customer through the return process, explains the steps involved, and generates a return label for them.

In the realm of business and customer service, ChatGPT enhances customer engagement, streamlines support operations, and delivers personalized experiences. It's important to note that while ChatGPT can handle a variety of customer inquiries, there might be cases where human intervention is necessary, especially for complex or sensitive issues. Additionally, businesses should ensure ethical use of customer data and provide clear communication regarding the involvement of AI in customer interactions.

Content Creation and Marketing

1. Blog Post and Article Generation:

   ChatGPT can assist content creators by generating blog posts and articles on various topics. It takes a given prompt, researches relevant information, and produces coherent and informative content. This is particularly useful for maintaining a consistent publishing schedule and scaling content production.

Example: A travel company needs to publish regular destination guides. ChatGPT generates a detailed guide to a specific location, including information about attractions, local cuisine, and travel tips.

2. Social Media Content:

Creating engaging and frequent social media content can be time-consuming. ChatGPT helps by generating posts, captions, and even replies to user comments. It tailors content to fit the platform's style and the brand's voice.

Example: A fashion brand wants to share daily outfit inspiration on Instagram. ChatGPT creates visually appealing captions that describe the outfits and provide styling tips.

3. SEO-Friendly Content:

ChatGPT can produce content optimized for search engines by incorporating relevant keywords and phrases naturally. This boosts the chances of content ranking higher in search results and attracting organic traffic.

Example: A company specializing in home improvement wants to create articles about DIY projects. ChatGPT ensures the articles include commonly searched terms related to home improvement and crafting.

4. Email Marketing Campaigns:

Crafting compelling email marketing campaigns is crucial for customer engagement. ChatGPT assists in writing email content that grabs the recipient's attention, promotes offers, and encourages conversions.

Example: An ecommerce business is launching a sale. ChatGPT helps create an email campaign that highlights the sale items, emphasizes the discounts, and includes persuasive call-to-action buttons.

5. Product Descriptions:

    When adding new products to an online store, writing unique and appealing product descriptions can be time-intensive. ChatGPT streamlines the process by generating product descriptions that highlight features and benefits.

    Example: A tech retailer introduces a new smartphone model. ChatGPT generates concise yet informative product descriptions that outline the phone's specifications, camera capabilities, and unique features.

6. Brand Messaging and Tone:

    Maintaining a consistent brand voice across different content platforms is essential. ChatGPT assists in creating content that aligns with the brand's messaging, values, and tone.

    Example: A fitness brand wants to communicate a motivational and empowering message. ChatGPT generates social media posts that inspire users to pursue their fitness goals and embrace a healthy lifestyle.

In the context of content creation and marketing, ChatGPT accelerates content generation, frees up time for strategizing, and ensures a steady flow of high-quality content. However, it's important to review and edit the content generated by ChatGPT to align with the brand's unique style and messaging. Additionally, human oversight ensures that the content accurately represents the brand's vision and resonates with the target audience.

Software Development and Tech Support

1. Code Assistance and Debugging:

    ChatGPT proves to be a valuable tool for developers seeking coding help. It can provide explanations of programming concepts, assist in debugging code, and even offer solutions to common coding problems.

Example: A developer encounters a syntax error in their code. ChatGPT helps identify the issue by analyzing the code snippet and suggesting corrections.

2. Explanation of Technical Concepts:

    Complex technical concepts can be challenging to grasp. ChatGPT acts as a knowledgeable companion, breaking down intricate ideas, algorithms, and theories into easily digestible explanations.

    Example: A computer science student struggles to understand the concept of recursion. ChatGPT provides a step-by-step explanation, clarifying the process and purpose of recursion.

3. Tech Troubleshooting and Problem-Solving:

    ChatGPT aids users in troubleshooting technical issues. It guides users through a series of questions to diagnose problems, suggests potential solutions, and provides instructions for resolution.

    Example: A user's printer isn't working. ChatGPT asks relevant questions about the printer's status, connectivity, and error messages. It then provides troubleshooting steps to resolve the issue.

4. Learning New Programming Languages:

    For developers venturing into new programming languages, ChatGPT offers guidance. It can generate sample code snippets, explain language syntax, and provide resources for learning.

    Example: A developer transitioning from Python to JavaScript seeks help with writing a function in JavaScript. ChatGPT provides a sample code snippet that accomplishes the desired task.

5. Documentation and API Usage:

    Navigating documentation and understanding APIs can be daunting. ChatGPT assists by explaining documentation, offering usage examples, and helping developers integrate APIs.

Example: A developer wants to integrate a payment gateway API into their ecommerce website. ChatGPT guides them through the API documentation and provides code snippets for integration.

6. Software Best Practices:

    ChatGPT can share insights into coding best practices, design patterns, and software architecture principles. It helps developers write cleaner, more efficient code.

    Example: A junior developer seeks advice on writing maintainable code. ChatGPT provides tips on modular programming, code commenting, and version control.

ChatGPT's applications in software development and tech support streamline the development process, enhance learning, and simplify problem-solving. However, developers should exercise caution and use their own judgment, especially in critical scenarios, as ChatGPT's solutions may not always account for context-specific considerations.

Data Entry and Analysis

Recent research ChatGPT's transformation into the code interpreter now called "Advanced Data Analysis" tool signifies a significant evolution in its capabilities. With this enhancement, it has become a powerful resource for data professionals and analysts, capable of not only understanding and generating code but also offering advanced insights into data analysis techniques, statistical modeling, data visualization, and more. This expanded functionality empowers users to extract deeper insights from their data, providing valuable assistance in a wide range of data-driven tasks and making it an invaluable asset in the field of data analytics:

1. Data Entry Assistance:

    ChatGPT assists in data entry tasks by transcribing handwritten or typed data, entering information into spreadsheets or databases, and organizing data according to specified formats.

    Example: A research team needs to digitize survey responses. ChatGPT transcribes the responses from paper forms into a digital spreadsheet.

2. Data Cleaning and Preprocessing:

    Before analysis, data often requires cleaning and preprocessing. ChatGPT helps identify and correct inconsistencies, missing values, and errors in the dataset.

    Example: An analyst is preparing a dataset for analysis. ChatGPT identifies and suggests corrections for duplicate entries and missing data points.

3. Basic Data Analysis and Visualization:

    ChatGPT performs simple data analysis tasks, such as calculating averages, generating charts, and summarizing trends. It aids in understanding basic insights from the data.

    Example: A marketing team wants to visualize sales data. ChatGPT generates bar charts and line graphs to illustrate sales trends over a specific time period.

4. Data Interpretation and Insights:

    ChatGPT assists in interpreting data findings, offering insights based on patterns and trends observed in the dataset. It provides explanations for significant findings.

    Example: An analyst notices a sudden drop in website traffic. ChatGPT suggests possible explanations, such as a recent algorithm change or a technical issue.

5. Comparative Analysis:

    ChatGPT aids in comparing datasets or different variables within a dataset. It helps identify correlations, differences, and relationships between data points.

    Example: A business wants to compare customer satisfaction ratings from two different product lines. ChatGPT calculates average satisfaction scores for each line and highlights differences.

6. Data Reporting and Summarization:

ChatGPT generates summaries and reports based on data analysis. It presents key findings, trends, and insights in a coherent and understandable format.

Example: An analyst needs to summarize a quarterly sales report. ChatGPT generates a concise report highlighting revenue trends, bestselling products, and regional performance.

ChatGPT's applications in data entry and analysis simplify data-related tasks, especially for basic analysis and organization. However, it's important to note that for complex data analysis, statistical modeling, and in-depth interpretation, involving data experts and analysts remains crucial for accurate insights and decision-making.

Healthcare and Medical Information

1. General Medical Information:

ChatGPT can provide general medical information to users seeking insights into symptoms, conditions, treatments, and preventive measures. It acts as a reliable source of introductory medical knowledge.

Example: A user experiences persistent headaches and seeks information about potential causes. ChatGPT offers explanations about various factors that could contribute to headaches and advises consulting a medical professional for accurate diagnosis.

2. Symptom Checker and Self-Assessment:

ChatGPT aids users in understanding their symptoms by asking targeted questions about their condition. It offers insights into potential causes and suggests whether seeking medical attention is advisable.

Example: A user describes symptoms like fever and body aches. ChatGPT engages in a symptom-checking conversation, suggests possible diagnoses like the flu, and advises rest and hydration.

3. Medication and Treatment Information:

For users curious about medication side effects, usage instructions, and potential interactions, ChatGPT provides relevant information based on its medical knowledge base.

Example: A user is prescribed a new medication and wants to know about possible side effects. ChatGPT outlines common side effects and advises the user to consult their healthcare provider if any adverse reactions occur.

4. Wellness Tips and Healthy Habits:

ChatGPT can offer general wellness advice, including tips on maintaining a healthy lifestyle, managing stress, and adopting preventive measures.

Example: A user asks about strategies for improving sleep quality. ChatGPT provides tips such as maintaining a consistent sleep schedule, creating a comfortable sleep environment, and limiting screen time before bed.

5. Explanation of Medical Terms:

Medical jargon can be intimidating for individuals without a medical background. ChatGPT simplifies medical terminology, explaining terms, acronyms, and abbreviations.

Example: A user comes across the term "hypertension" and is unsure about its meaning. ChatGPT explains that it refers to high blood pressure and provides a brief overview of its implications.

6. Preparing for Medical Appointments:

ChatGPT helps users prepare for medical appointments by suggesting questions to ask healthcare providers, highlighting important information to share, and offering tips for effective communication.

Example: A user is scheduled for a doctor's appointment regarding a chronic condition. ChatGPT provides a list of questions to ask the doctor, ensuring the user gathers all necessary information.

ChatGPT's role in healthcare offers accessible information and guidance, especially for preliminary understanding and nonurgent queries. However, it's crucial to emphasize that ChatGPT should never replace professional medical advice. Users should always consult qualified healthcare professionals for accurate diagnoses and treatment recommendations.

Market Research and Analysis

1. Survey Analysis and Summarization:

   ChatGPT can analyze survey responses and summarize key findings. It assists researchers by identifying common trends, sentiments, and patterns within large sets of survey data.

   Example: A company conducts a customer satisfaction survey. ChatGPT reviews the survey results, highlights areas with the highest satisfaction ratings, and identifies recurring concerns.

2. Customer Feedback Insights:

   Businesses receive vast amounts of customer feedback across various platforms. ChatGPT aids in extracting insights from these feedback channels, categorizing comments, and identifying emerging trends.

   Example: An ecommerce retailer wants to understand customer sentiments from product reviews. ChatGPT categorizes feedback into positive, negative, and neutral sentiments, providing an overview of customer opinions.

3. Competitor Analysis:

    ChatGPT assists businesses in analyzing their competitors by collecting information from various sources and summarizing their strengths, weaknesses, market positioning, and strategies.

    Example: A tech startup wants to evaluate its competitors in the smartphone market. ChatGPT compiles information about competitors' features, pricing, and user reviews, offering a comprehensive analysis.

4. Trend Identification and Forecasting:

    ChatGPT can analyze market trends by processing data from social media, news articles, and industry reports. It identifies emerging trends and patterns that can guide strategic decision-making.

    Example: A fashion brand wants to predict the next season's popular clothing styles. ChatGPT analyzes social media conversations and fashion blogs to forecast upcoming trends.

5. Consumer Behavior Analysis:

    ChatGPT assists in understanding consumer behavior by analyzing purchasing patterns, preferences, and buying motivations. It provides insights that inform marketing campaigns and product development.

    Example: An online retailer wants to understand why certain products are popular during specific seasons. ChatGPT analyzes purchasing data and identifies trends in consumer behavior.

6. Market Segment Profiling:

    ChatGPT helps businesses profile different market segments based on demographic, geographic, and psychographic factors. It aids in tailoring marketing strategies to specific audience segments.

Example: An electronics manufacturer wants to target a specific demographic for a new product launch. ChatGPT creates profiles of potential customers, outlining their preferences and interests.

ChatGPT's applications in market research and analysis streamline data processing, offer actionable insights, and enable businesses to make informed decisions. However, human expertise remains essential to interpret and contextualize results, ensuring that business strategies are grounded in a well-rounded understanding of market dynamics.

Creative Writing and Storytelling

1. Idea Generation and Brainstorming:

   ChatGPT becomes a creative collaborator, assisting writers in generating ideas for stories, articles, blog posts, and creative projects. It sparks creativity by suggesting plotlines, characters, settings, and themes.

   Example: An author is stuck while brainstorming ideas for a new novel. ChatGPT proposes a unique concept involving time travel and alternate realities, reigniting the author's creative process.

2. Plot Development and Story Outlining:

   ChatGPT helps writers structure their stories by providing guidance on plot development. It assists in creating story arcs, building suspense, and mapping out the sequence of events.

   Example: A screenwriter wants to outline a compelling TV series pilot. ChatGPT assists in crafting the pilot episode's plot, introducing characters, and setting up future storylines.

3. Character Creation and Development:

   Crafting engaging characters is crucial to storytelling. ChatGPT aids writers in developing well-rounded characters by suggesting personality traits, backstories, motivations, and character arcs.

Example: A fantasy writer is creating a new protagonist. ChatGPT suggests a complex backstory involving a tragic event and a hidden magical ability, adding depth to the character.

4. Dialogue Writing:

Natural and engaging dialogue is integral to storytelling. ChatGPT helps writers create authentic dialogues by suggesting conversational lines, emotional nuances, and interactions between characters.

Example: A playwright is working on a dramatic scene. ChatGPT offers lines of dialogue that convey tension and conflict between characters, enhancing the scene's impact.

5. Worldbuilding and Setting Descriptions:

For immersive storytelling, vivid worldbuilding and descriptive settings are essential. ChatGPT assists writers in creating richly detailed settings and evocative descriptions.

Example: A science fiction author wants to describe an alien planet. ChatGPT provides sensory details about the planet's unique flora, fauna, and atmosphere, painting a vivid picture.

6. Creative Prompts and Writing Exercises:

ChatGPT offers creative prompts and writing exercises to overcome writer's block and stimulate the imagination. It provides starting points for short stories, poems, and creative experiments.

Example: A poet is seeking inspiration for a new poem. ChatGPT provides a thought-provoking prompt about the beauty of nature, inspiring the poet to craft a descriptive piece.

ChatGPT's applications in creative writing and storytelling empower writers to overcome challenges, explore new ideas, and breathe life into their narratives. While it aids in the creative process, human judgment and editing remain crucial for ensuring narrative coherence, emotional resonance, and the writer's unique voice.

Education and Learning

1. Virtual Tutoring and Concept Explanation:

    ChatGPT serves as a virtual tutor, assisting students in understanding complex concepts. It explains academic subjects, breaks down theories, and offers step-by-step solutions to problems.

    Example: A high school student struggles with calculus. ChatGPT provides explanations for calculus principles and helps solve practice problems, aiding the student's understanding.

2. Homework and Assignment Help:

    ChatGPT aids students in completing homework and assignments by providing guidance, suggesting approaches, and answering questions related to the tasks.

    Example: A student has to write an essay on a historical event. ChatGPT offers research suggestions, outlines key points, and provides insights to structure the essay effectively.

3. Language Learning and Practice:

    ChatGPT becomes a language learning companion, engaging learners in conversations, correcting sentences, and suggesting vocabulary words to enhance language proficiency.

    Example: A language learner wants to practice Spanish. ChatGPT engages in a conversation, corrects grammar errors, and introduces new vocabulary in context.

4. Study Resource Generation:

    ChatGPT assists students by generating study resources such as flashcards, summaries, and practice questions. It condenses lengthy material and helps students review effectively.

    Example: A student prepares for a history exam. ChatGPT generates concise summaries of key historical events, aiding the student's last-minute review.

5. Research Assistance:

    For research projects, ChatGPT aids students in finding relevant sources, formulating research questions, and organizing information to create well-structured papers.

    Example: A college student is conducting research on climate change. ChatGPT suggests reputable sources, helps refine research questions, and outlines a research paper structure.

6. Exploring New Topics and Curiosities:

    ChatGPT encourages curiosity-driven learning by providing explanations on a wide range of topics. It satisfies learners' queries and stimulates further exploration.

    Example: A curious learner wants to understand the basics of quantum physics. ChatGPT offers an introductory explanation, demystifying complex concepts.

ChatGPT's applications in education and learning extend beyond traditional classrooms, offering personalized assistance, fostering self-directed learning, and aiding students in their academic journey. While ChatGPT enhances learning experiences, educators' guidance, curricular structure, and critical thinking development remain essential components of effective education.

Legal and Compliance

1. Legal Research and Case Law Analysis:

    ChatGPT assists legal professionals by conducting legal research and summarizing case law. It extracts relevant information from legal databases, helping lawyers build stronger arguments and make informed decisions.

    Example: A lawyer is preparing a case involving intellectual property rights. ChatGPT compiles relevant case law examples, aiding the lawyer's understanding of precedent.

2. Drafting Legal Documents:

    ChatGPT aids in drafting legal documents such as contracts, agreements, and letters. It generates templates, provides guidance on language and structure, and ensures documents adhere to legal norms.

    Example: An entrepreneur needs a nondisclosure agreement. ChatGPT helps create a comprehensive agreement, including confidentiality clauses and legal terminology.

3. Legal Definitions and Explanations:

    Legal terminology can be intricate for nonlegal professionals. ChatGPT simplifies legal concepts by providing definitions, explanations, and context for various legal terms.

    Example: A business owner encounters the term "tort." ChatGPT explains the concept of tort law, its types, and implications for business operations.

4. Compliance Guidelines and Regulations:

    ChatGPT assists businesses in understanding and adhering to legal regulations and compliance standards. It offers explanations of regulatory requirements and suggests steps for compliance.

    Example: A company wants to ensure compliance with data protection regulations. ChatGPT outlines the key provisions of relevant data privacy laws and provides recommendations for compliance.

5. Legal Advice for Common Issues:

    For everyday legal questions and concerns, ChatGPT offers preliminary legal advice and guidance. It addresses queries related to contracts, employment law, liability, and more.

    Example: A small business owner is uncertain about employee termination procedures. ChatGPT explains the legal steps involved in compliantly terminating an employee.

6. Intellectual Property Guidance:

ChatGPT assists in navigating intellectual property matters by providing insights into copyright, trademarks, and patents. It explains the process of registering and protecting intellectual property.

Example: An artist wants to protect their artwork from unauthorized use. ChatGPT explains the basics of copyright law, including how to register their work.

ChatGPT's applications in legal and compliance streamline legal research, simplify documentation processes, and offer preliminary guidance. However, it's important to note that ChatGPT's responses should not replace professional legal advice. Legal professionals should be consulted for complex legal matters and critical decisions.

HR and Recruitment

1. Candidate Screening and Initial Interviews:

ChatGPT assists HR professionals in conducting preliminary candidate screenings. It engages with applicants, asks relevant questions, and evaluates responses to shortlist candidates for further evaluation.

Example: An HR manager needs to screen a high volume of job applications. ChatGPT conducts brief interviews with applicants, asking about their qualifications and experience.

2. Job Description Crafting:

Crafting compelling job descriptions is essential for attracting suitable candidates. ChatGPT assists in creating detailed and engaging job postings that highlight responsibilities, qualifications, and company culture.

Example: A company is hiring a social media manager. ChatGPT generates a job description that effectively communicates the role's expectations and the company's brand.

3. Employee Onboarding Support:

    ChatGPT aids in employee onboarding by providing information about company policies, benefits, and the onboarding process. It answers new hires' questions and ensures a smooth transition.

    Example: A new employee wants to know more about the company's vacation policy. ChatGPT provides an overview of the policy and how to request time off.

4. Training and Development Assistance:

    HR professionals can use ChatGPT to offer training resources and development opportunities. It recommends online courses, workshops, and skill-building activities based on employees' career goals.

    Example: An employee expresses interest in improving their project management skills. ChatGPT suggests relevant courses and resources for professional development.

5. Employee Assistance and Policy Clarification:

    ChatGPT assists employees in understanding company policies, benefits, and HR procedures. It provides information about leave policies, grievance procedures, and more.

    Example: An employee wants to know the procedure for reporting workplace harassment. ChatGPT explains the steps to follow and emphasizes the importance of reporting.

6. Interview Preparation and Tips:

    For job seekers, ChatGPT offers interview preparation guidance. It suggests common interview questions, provides tips for effective responses, and offers insights into interview etiquette.

    Example: A job applicant is nervous about an upcoming interview. ChatGPT provides advice on how to prepare, answer questions confidently, and make a positive impression.

ChatGPT's applications in HR and recruitment optimize hiring processes, enhance candidate experiences, and streamline communication between HR professionals and employees. While ChatGPT can support various tasks, it's important to note that human involvement remains essential for nuanced decision-making, evaluating soft skills, and addressing complex HR matters.

Personal Assistant and Productivity

1. Task Management and Reminders:

    ChatGPT acts as a virtual task manager, helping users organize their to-do lists, set reminders for appointments, and manage deadlines for tasks and projects.

    Example: A user schedules a meeting and asks ChatGPT to remind them 15 minutes before the meeting starts.

2. Calendar Coordination:

    ChatGPT assists in scheduling and coordinating events. It checks availability, proposes suitable meeting times, and helps users schedule appointments.

    Example: A professional wants to set up a virtual meeting with colleagues across different time zones. ChatGPT suggests optimal meeting times that accommodate everyone's schedules.

3. Information Retrieval:

    ChatGPT quickly retrieves information from the Web or databases, saving users time in searching for facts, figures, definitions, or historical data.

    Example: A student needs information for a research paper. ChatGPT retrieves relevant articles, statistics, and sources on the chosen topic.

4. Note-Taking and Summarization:

    ChatGPT assists in taking notes during meetings, classes, or conferences. It can also summarize lengthy documents, distilling key points for easy reference.

    Example: A user attends a conference and asks ChatGPT to take notes. ChatGPT creates a concise summary of the conference sessions.

5. Language Translation on the Go:

    ChatGPT serves as a language translator, aiding users in real-time translation of conversations, text, or content from one language to another.

    Example: A traveler needs assistance with translating street signs and menus while exploring a foreign country. ChatGPT provides instant translations.

6. Personalized Recommendations:

    ChatGPT suggests books, movies, music, restaurants, and more based on users' preferences. It can help users discover new content or make decisions.

    Example: A user asks ChatGPT for book recommendations in the mystery genre. ChatGPT provides a list of highly rated mystery novels.

7. Fitness and Wellness Assistance:

    ChatGPT offers workout routines, nutrition advice, and wellness tips. It helps users set fitness goals and suggests exercises based on their preferences.

    Example: A user wants to start a home workout routine. ChatGPT designs a personalized workout plan with different exercises and intensity levels.

ChatGPT's applications as a personal assistant and productivity tool streamline daily tasks, enhance organization, and provide convenient access to information. However, while ChatGPT can handle various tasks, human judgment and decision-making are crucial, especially in scenarios requiring complex reasoning or subjective evaluations.

Examples

Till now we saw the use case by domain using ChatGPT. Now let us take a few use-case examples and ask ChatGPT:

1. Domain—HR and Recruitment:

   For Bangalore location, let us ask to create a job description for a data scientist who has a minimum experience of three years and must have Python and PySpark hands-on experience in healthcare domain.

   Figure 9-1 shows ChatGPT's response.

Job Title: Data Scientist

Location: Bangalore, India

Employment Type: Full-time

Experience Level: Mid-Level (Minimum 3 years of experience)

Company Overview:

[Company Name] is a pioneering healthcare technology company dedicated to transforming the industry through data-driven solutions. We are seeking an experienced Data Scientist to join our Bangalore team. This role presents an exciting opportunity to work on cutting-edge healthcare projects, utilizing your hands-on expertise in Python and PySpark for advanced analytics and predictive modeling.

Key Responsibilities:

1. **Data Analysis and Modeling:** Apply advanced statistical and machine learning techniques to analyze extensive healthcare datasets. Develop predictive models to unearth critical insights and trends.
2. **Data Preprocessing:** Meticulously clean, preprocess, and transform raw healthcare data into meaningful formats. Address missing data, outliers, and ensure data integrity.
3. **Python and PySpark Proficiency:** Harness the power of Python and PySpark for data manipulation, analysis, and modeling. Construct and optimize PySpark pipelines for large-scale data processing.
4. **Healthcare Domain Expertise:** Utilize your deep knowledge of the healthcare domain to navigate complex datasets, including electronic health records (EHR), medical claims, and clinical data.
5. **Feature Engineering:** Engineer relevant features from healthcare data to enhance model accuracy and performance.
6. **Model Deployment:** Collaborate with the engineering team to deploy machine learning models into production, enabling real-time decision support.

Figure 9-1. *ChatGPT's response for example 1*

7. **Cross-Functional Collaboration:** Forge strong partnerships with cross-functional teams, including data engineers, clinicians, and product managers, to define project objectives and deliver actionable insights.

8. **Continuous Learning:** Stay at the forefront of data science, machine learning, and healthcare trends, infusing innovative solutions into our projects.

Qualifications:

* Bachelor's or Master's degree in Computer Science, Statistics, Mathematics, or a related field.
* Minimum of 3 years of proven experience as a Data Scientist, with a strong track record in the healthcare domain.
* Proficiency in Python for data analysis and machine learning libraries (e.g., NumPy, pandas, scikit-learn).
* Hands-on experience with PySpark for big data processing and distributed computing.
* Expertise in statistical analysis, hypothesis testing, and machine learning algorithms.
* Familiarity with data visualization tools (e.g., Matplotlib, Seaborn, Tableau).
* Strong analytical and problem-solving abilities, coupled with adaptability in a dynamic environment.
* Exceptional communication skills to convey complex findings and insights effectively to non-technical stakeholders.
* Knowledge of healthcare data standards and privacy regulations (e.g., HIPAA) is a significant advantage.

How to Apply:
If you are a passionate and experienced Data Scientist who is excited about making a meaningful impact in healthcare through data science, we encourage you to submit your resume and a comprehensive cover letter. Please include any relevant project portfolios or GitHub repositories.

[Company Name] is an equal opportunity employer committed to fostering diversity and inclusion in the workplace. We welcome applications from candidates of all backgrounds and experiences.

Figure 9-1. *(continued)*

2. Domain—Software Development and Tech Support:

Explain the following code snippet:

```python
def binary_search(arr, low, high, x):
    if high >= low:
        mid = (high + low) // 2
        if arr[mid] == x:
            return mid
        elif arr[mid] > x:
            return binary_search(arr, low, mid - 1, x)
        else:
            return binary_search(arr, mid + 1, high, x)
    else:
        return -1
```

Figure 9-2 shows ChatGPT's response.

The provided code snippet is an implementation of the binary search algorithm in Python. Binary search is a highly efficient algorithm used to find the position (index) of a target element (`x`) within a sorted array (`arr`).

Here's a breakdown of how the code works:

1. **Function Definition:** The code defines a function named `binary_search` that takes four arguments:
   * `arr`: This is the sorted array in which the algorithm searches for the target element.
   * `low`: This represents the lower index of the current search range within the array.
   * `high`: This represents the upper index of the current search range within the array.
   * `x`: This is the target element that we want to find within the array.
2. **Initial Check:** The function begins by checking if `high` is greater than or equal to `low`. This condition ensures that the search is valid within the current search range. If `high` is less than `low`, it means the target element is not in the array, and the function returns `-1` to indicate that the element was not found.
3. **Calculate Middle Index:** Inside the valid search range, the code calculates the middle index `mid` using integer division (`//`) of the sum of `high` and `low` by 2. This gives the index of the element in the middle of the current search range.
4. **Comparison:** The code then compares the element at the middle index `arr[mid]` with the target element `x`:
   * If they are equal (`arr[mid] == x`), it means the target element has been found, and the function returns `mid`, which is the index where `x` is located in the array.
   * If `arr[mid]` is greater than `x`, it means that `x` must be in the left half of the current search range. Therefore, the function is recursively called with the new search range `(low, mid - 1)`.
   * If `arr[mid]` is less than `x`, it means that `x` must be in the right half of the current search range. The function is recursively called with the new search range `(mid + 1, high)`.
5. **Base Case:** If the target element is not found within the current search range (when `high` becomes less than `low`), the function returns `-1` to indicate that the element was not found.

Figure 9-2. *ChatGPT's response for example 2*

3. Domain—Education and Learning:

Translate how are you in Hindi.

Figure 9-3 shows ChatGPT's response.

 "How are you" in Hindi is translated as "कैसे हो आप?" (pronounced as "Kaise ho aap?")

Figure 9-3. *ChatGPT's response for example 3*

4. Question Answering:

Who is the father of computer?

Figure 9-4 shows ChatGPT's output.

 The title "Father of the Computer" is often attributed to Charles Babbage, an English mathematician, philosopher, inventor, and mechanical engineer. Babbage is known for his pioneering work in the field of computing during the 19th century.

He conceptualized the idea of a mechanical general-purpose computer, which he called the "Analytical Engine." Babbage's designs for the Analytical Engine included features like a central processing unit (CPU), memory, punch cards for input, and a printer for output. Although the Analytical Engine was never fully built during his lifetime due to various challenges, his ideas laid the foundation for modern computing.

Babbage's work on the Difference Engine, an earlier mechanical calculator designed to compute polynomial functions, is also considered significant in the history of computing.

Figure 9-4. *ChatGPT's response for example 4*

Conclusion

The versatile applications of ChatGPT across various domains showcase its transformative potential. Whether it's enhancing customer service interactions, streamlining content creation and marketing efforts, facilitating language and communication tasks, empowering software development and tech support, revolutionizing healthcare and medical information management, or driving market research and analysis, ChatGPT consistently proves its adaptability and utility. Additionally, its proficiency in creative writing, education, legal compliance, HR functions, and data analysis further underscores its value across diverse sectors. With its ability to comprehend, generate, and assist in decision-making, ChatGPT emerges as a remarkable tool that continues to redefine how we harness the power of AI for real-world solutions in today's dynamic landscape.

Index

© Akshay Kulkarni, Adarsha Shivananda, Anoosh Kulkarni, Dilip Gudivada 2023
A. Kulkarni et al., *Applied Generative AI for Beginners*, https://doi.org/10.1007/978-1-4842-9994-4

V, W, X, Y

Z

Printed in the United States
by Baker & Taylor Publisher Services